Field Guide to

EDIBLE MUSHROOMS

of BRITAIN AND EUROPE

This book is dedicated to Valerie for her help and support to me that have made our many changes of direction possible.

Publisher's note: author Peter Jordan sadly passed away in November 2008. We hope that his book will continue to inspire mushroom-hunters for many years to come.

This edition published in 2010 by New Holland Publishers (UK) Ltd
London • Cape Town • Sydney • Auckland
www.newhollandpublishers.com

Garfield House, 86–88 Edgware Road, London W2 2EA, United Kingdom

80 McKenzie Street, Cape Town 8001, South Africa

Unit 1, 66 Gibbes Street, Chatswood, NSW 2067, Australia

218 Lake Road, Northcote, Auckland, New Zealand

4 6 8 10 9 7 5 3

ISBN 978 1 84773 746 5

Publisher: Simon Papps
Editor: Gareth Jones
Assistant Editor: Kate Parker
Design Concept: Alan Marshall
Designer: Peter Davies/D&N Publishing
Photographers: Peter Henley & Peter Jordan
Artist: Peter Henley
Cartoonist: Tony Hall
Production: Joan Woodroffe
Indexer: Angie Hipkin

Reproduction by Modern Age Repro House Ltd, Hong Kong
Printed and bound in Malaysia by Times Offset (M) Sdn Bhd

DISCLAIMER:
The publishers and the author cannot accept responsibility for any identification of any mushroom made by the users of this guide, nor for any effects resulting from eating any wild mushroom. Although many species are edible for the majority of people, some species can cause allergic reactions or illness to a number of people: this is totally unpredictable.

Field Guide to

EDIBLE MUSHROOMS

of BRITAIN AND EUROPE

PETER JORDAN

NEW HOLLAND

Contents

FROM POUNDS TO PINTS TO PENNY BUNS (*Boletus Edulis*)

The Penny Bun (*Boletus edulis*).

I have been collecting wild mushrooms for over 60 years. I was introduced to the hunt by my grandfather, a Norfolk farmer, who delighted in collecting from the wild. I was brought up on the farm during the war years when everything was in short supply, so it was no surprise that free food was often on our table. The art of collecting faded in the post-war years, but these days more and more restaurants are serving wild mushrooms. And so, as a result, there is a renewed interest in collecting the more interesting fungi.

Although a country boy at heart, after my schooling, I went into banking, spending 28 years travelling the country in this career, and ending up as a bank manager in Cambridge. However, I never lost my interest in collecting wild mushrooms. I also never forgot my grandfather's wise advice. Be sure you've identified your mushroom correctly – if in doubt leave it out.

By 1984, I'd had enough of banks and offices. With the support of my wife Valerie and our children, we embarked on a new venture: running a country pub in my native Norfolk. We spent 10 happy and very busy years in the small village of Burnham. As we put more and more mushroom dishes on our menu we soon became known as 'the mushroom pub'. Of course, when we weren't serving the public, we often went on mushroom hunts in the local countryside.

After the lease on 'the mushroom pub' expired, I began promoting the wonder of wild mushrooms by writing about them, making a video, organising and leading mushroom forays, and helping at cookery schools at home and abroad.

Since then, I've worked with Gary Rhodes, Brian Turner and Hugh Fearnley-Whittingstall. We have also moved to Southwold in Suffolk. 'Poppy Cottage' has a wonderful view over the common and golf course, and I can even collect mushrooms from my front garden. This is now the home of the *Tasty Mushroom Partnership*, promoting the enjoyment of mushrooms everywhere.

Peter and Valerie on the fungi trail.

AN INTRODUCTION

The terms 'mushroom' and 'toadstool' often cause people confusion. Mushrooms and toadstools are all fungi. Some are edible, but others are poisonous, even deadly. In this book, we help you select the best of the edible varieties; we also warn you about the lookalikes – especially those that are poisonous or deadly.

Many books about mushrooms have been published over the years, but people are often confused by the sheer number of species to be identified. And so, in this book, we concentrate on the ones you need to know for the cooking pot, while clearly identifying the ones you must be sure to avoid.

However, this book is a guide that should be used in connection with a comprehensive mycological guide with detailed identification data. There are quite a number of good reference books available, many of which you can find in the *Further Reading* section on page 152. There have been a number of species name changes in recent years, so check carefully.

Fungi, which include yeasts and mould, as well as mushrooms and toadstools, are fascinating. Because they don't contain the green pigment chlorophyll or use the process of photosynthesis, they don't need daylight in order to grow. As a result, most fungi develop during the hours of darkness, and their growth is linked to the phases of the moon.

In this book we look at two main groups of fungi: the Ascomycetes and the Basidiomycetes.

Gastronomy vs. mycology.

Black gold – a Black Summer Truffle (*Tuber aestivum*).

The Ascomycetes are spore-shooters: they rely on the wind to spread their spores, so as to produce more fungi. An important family in this group is the Morels, which grow in the spring. When you find one, you can check along the prevailing wind lane, where the spores are blown, to find more of them.

The Basidiomycetes are a very large group, which also rely predominantly on the wind to spread their spores. They grow from underground 'mycelial' threads. In the right conditions, the threads can bunch together to form fruit. These mushrooms shed spores from their gills (or from their pores in the case of Boletus mushrooms), which can form a new mycelium.

I prefer to cut wild mushrooms (unless it is essential to lift them for positive identification). This cutting causes minimal damage to the underground mycelial branches, but most damage is done by compacting of soil – so be careful where you tread!

Some fungi don't even appear above the ground, relying instead on animals finding them, in order to spread their spores. The truffle is a classic example. Animals root for truffles, eat them, and spread the spores in their droppings. Dogs are now used rather than pigs, as they don't eat the truffles they find. They are also easier to transport and handle, helping hunters to keep their locations secret.

In this book you will find information about when and where to hunt mushrooms. We explain how to hunt for them in a sustainable way, preserving the environment and encouraging new growth in the next season. We give you all the details you need about the different parts of each fungus, which are important in identifying them correctly. We also tell you how to store the mushrooms you've picked if you aren't going to eat them straight away: mushrooms don't last long, so easy storage methods for your finds are useful to know.

I believe that ensuring safety and sustainability are the most important elements of mushroom hunting, and there is nothing like going out with an expert on a foray to increase your skills. Good reference books are essential but they cannot make up for practical experience in the field.

MUSHROOMS IN HISTORY

Human beings have been fascinated by wild mushrooms for centuries. While many historical facts have sadly been lost, we do know that mushrooms formed a much-prized part of the diet of Roman noblemen, and latterly that of the Victorians. The Roman influence is still visible in the many Latin scientific names that are used worldwide today.

On the forays I lead, people often ask me how it was discovered which mushrooms could be safely eaten. The answer has to be 'trial and error' – the error sometimes being a fatal one for the unlucky eater. (This is true of most plants, of course. In America they have even put up a statue of the volunteer who first ate a tomato to find out if it was poisonous.)

The Romans were fond of Caesar's Mushroom (*Amanita caesarea*) (see page 34) and it was often made a special feature in lavish banquets. There was a snag, however: if a Death Cap (*Amanita phalloides*) (see page 118) or a Destroying Angel (*Amanita virosa*) (see page 120) was introduced to the dish, it was goodbye to Caesar (and his food taster). The poison takes at least 6–14 hours to manifest its grim symptoms; death follows about seven days later. This was a method of assassination employed by several particularly ambitious Roman statesmen, and is well documented.

Caesar's Mushroom (*Amanita caesarea*).

The Death Cap (*Amanita phalloides*).

For centuries, people in the British Isles have regarded fungi with great suspicion. Mushrooms have often been associated with witchcraft, disease and death; the fact that they only grow in the hours of darkness and that they are affected by the different phases of the moon only adds to the suspicion, which has lingered to this day.

The ancient Druids, too, have influenced our wary attitude to mushrooms. Under Druidic law, only the priests themselves were allowed to gather mushrooms. This was because they used some of the hallucinogenic varieties in their ceremonies, to induce a state of trance and euphoria, adding to the power and mystery of the rites.

The Vikings had an interesting use for one particular mushroom, the dangerous Fly Agaric (*Amanita muscaria*) (see page 117). Dried *A. muscaria* caps were fed to the Berserkers in the Viking longships just before battle. As a result the warriors first went into a trance, then experienced a dramatic increase in their physical strength for battle, in which they were famously frenzied, with sufficient energy left over for the rape and pillage afterwards. The Vikings got the idea from observing what happened to reindeer, which enjoyed eating *Amanita muscaria*. Its intoxicating effects gave the animals obvious pleasure – and the illusion that they could fly. I have myself been astounded by the sight of several spaced-out reindeer in Svalbard: they certainly behaved as though they could become airborne – inspiration, perhaps, for that wonderful Christmas song *Rudolf the Red-nosed Reindeer*!

The Fly Agaric is, in fact, very dangerous, not least because you can never tell what level of toxin any individual cap contains. There was a case recently in Birmingham of a man who had eaten a Fly Agaric and was found in the forest in a trance. He was quickly taken by helicopter to the intensive care unit at a nearby hospital. I was telephoned for advice, and told them to put the patient in a straitjacket and strap him to his bed. Not surprisingly, the ward sister was alarmed by such extreme measures but I assured her it

Fly Agaric (*Amanita muscaria*).

would take at least five people to restrain him when he came out of the trance. Another phone call two hours later confirmed my fears: the man had actually burst the straitjacket. Thank goodness that the bed restraints had worked.

The other problem with Fly Agaric is that the toxins stay in your system for a very long time. This means you continue to have sudden episodes of trance and delusion, which can include that feeling of being able to fly. Not good if you are driving on the M25 at the time! Not surprisingly, the best thing to do is leave *Amanita muscaria* well alone.

Destroying Angel (*Amanita virosa*).

Wild mushrooms came into their own during the two world wars. Food was scarce throughout Europe, so wild food gathering became essential. People in rural communities, especially in central and eastern Europe, revived or gained the knowledge they now needed about mushrooms in the wild. That knowledge has been passed down to their descendants, whose dinner tables benefit today. The French and the Italians, of course, already knew about the culinary value of wild mushrooms, which have been an important part of their cooking for centuries.

Increasingly, nowadays, the gourmet delights of wild mushrooms are rising to a new level of esteem. This is partly because people travel so much more than they did. It is also because of the prominence of television chefs and cookery programmes, highlighting the versatility and superb flavours of these fruits of the countryside.

I recently travelled with my wife Valerie to Australia, where I was fascinated by a particular aspect of Aboriginal bush tucker. It certainly included a number of wild mushrooms, but to my great surprise the Aboriginals ignored the Chicken of the

Woods or Sulphur Polypore (*Laetiporus sulphureus*) (see page 72), because they did not know it was edible. I persuaded some of them to try it out and they enjoyed it. When we got home, there waiting for me was my copy of the *British Mycological Journal*, containing a report from a research team in North Borneo. The team had discovered that the local Indians

Chicken of the Woods (*Laetiporus sulphureus*).

did not eat the Chicken of the Woods – it appeared not to be part of their culture. So the team and the Indians enjoyed a mushroom feast together, including Chicken of the Woods. Mushroom history had been made, independently, in the East and the Antipodes! **Please note, however, that you must be careful with this species, as around 10 per cent of the population can have an allergic reaction to it.**

Mushrooms also have had a part to play in medicine. The most famous and best-documented example, of course, is that of Alexander Fleming's discovery of penicillin: a marvellous medical breakthrough. Mushrooms of the Puff-ball family have traditionally been used to staunch bleeding, and they are sometimes still used today on horses. These mushrooms also have some antiseptic qualities. When the Giant Puff-ball (*Calvatia gigantea*) (see page 49) was shedding its spores it was used by beekeepers to subdue the hive. The smaller Var. Puff-ball (*Lycoperdon perlatum*) (see page 80) was used (again in the spore-shooting stage) by dentists as a painkiller. I have to say I'd

A Puff-ball in action.

rather have an injection. Discoveries of such mushroom virtues are still being made today, especially in treatments for cancer. The Hen of the Woods (*Grifola frondosa*) (see page 61) is proving successful with stomach cancer.

Hen of the Woods (*Grifola frondosa*).

Another centuries-old use of fungi is in the production of dyes. A friend of mine on the Isle of Harris still uses a number of fungi to dye her wool naturally for the famous Harris Tweed. She procures a great array of colours from them, ranging from orange to green. My tweed mushroom breeches are a notable testament to the success of my friend and these fungal dyes.

Natural fungal dyes.

THE DIFFERENT PARTS OF A MUSHROOM

The Blusher (*Amanita rubescens*).

The very name 'mushroom' causes some confusion. Some people only use the term for Field Mushrooms (*Agaricus campestris*) (see page 30), calling most of the rest 'toadstools'. The whole group are in fact Fungi. For the purposes of this book they will be referred to as mushrooms. The term 'fungi' covers an enormous range of organisms, including moulds, yeasts, and a host of microscopic examples.

Mushrooms do a good job, by and large, of breaking down dead or dying material into food for trees and plants. Most are symbiotic, producing nutrients that are used by the trees, on or near to which they grow. Some mushrooms, however, are parasitic and kill their host, such as the Honey Fungus (*Armillaria mellea*) (see page 37), which has laid acres of woodland to waste.

To aid collecting, it is important that you recognise the various types of mushrooms, as well as their various parts: this will help you to make clear and accurate identifications.

The two main groups featured in this book are the Ascomycetes and the Basidiomycetes. Ascomycetes are spore shooters, such as Morels, which rely on the wind to spread their spores. The truffle, which relies on animals to eat it to spread its spores, is also in this group. The Basidiomycetes include the more traditionally-shaped mushrooms, such as the Field Mushroom (*Agaricus campestris*) (see page 30) and the Horse Mushroom (*Agaricus arvensis*) (see page 28). Then, there is another group of more unusual mushrooms featured in this book: the Aphyllophorales, which are also members of the Basidiomycetes. The most-collected of these is the Cauliflower Fungus (*Sparassis crispa*) (see page 104).

Within the main group of Basidiomycetes there are two distinct types: those with **gills** (like the Field Mushroom) and those with **pores** – the Boletus mushrooms, which include the Cep (*Boletus edulis*) (see page 44).

The mushroom's **spores** – the equivalent of seeds – are carried by **tubes** (which open via the pores), or by the gills. The spores are ready for release when a mushroom is mature, and that is why we only collect mature specimens: as we pick them and place them in our mushroom baskets, we are also spreading the spores, which will grow into next year's crop.

Other parts important to recognise: the **cap**, the **stem**, and the **ring** if there is one. A mushroom's **shape**, **colour** and **aroma** are also important guides to identification. The Agaric type of mushroom often has a **partial veil**, which covers only the gills. As the mushroom grows out of

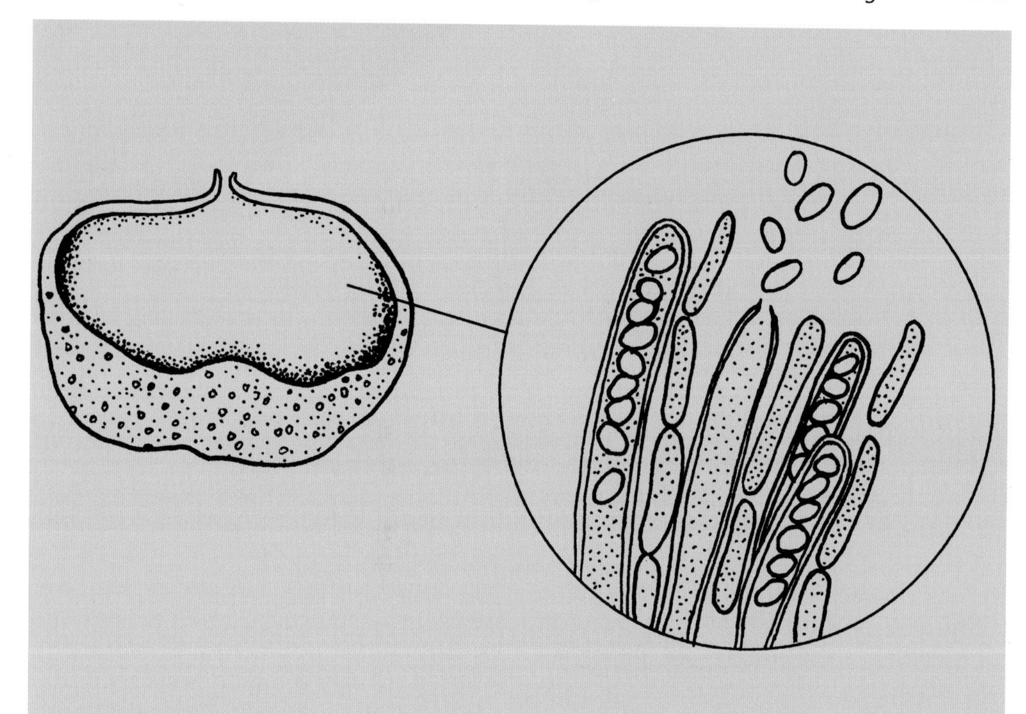

The fruiting body of the Ascomycetes (spore-shooting) group of mushrooms, which includes the Morels and Puff-balls. These use the wind to spread their spores.

its **volval sac**, the veil ruptures to leave a ring on the stem. The **universal veil**, which is more or less restricted to the Amanita species, encloses the whole mushroom. When a veil ruptures, it sometimes leaves remnants on the cap. The most recognisable volval sac is that found on the Death Cap (*Amanita phalloides*) (see page 118), which looks like a hatching egg in its early stages.

A spore print example.

Some of the collectors I know like to take spore prints, not only for their looks but also to help with identification of some of the more difficult species. While these can be beautiful, they also give you some idea how many spores an individual cap contains. Pick a mature mushroom and place it on a white card (or on a dark card if the spores are likely to be pale). Then leave it in a warm place for two days. You should then find you have a good print.

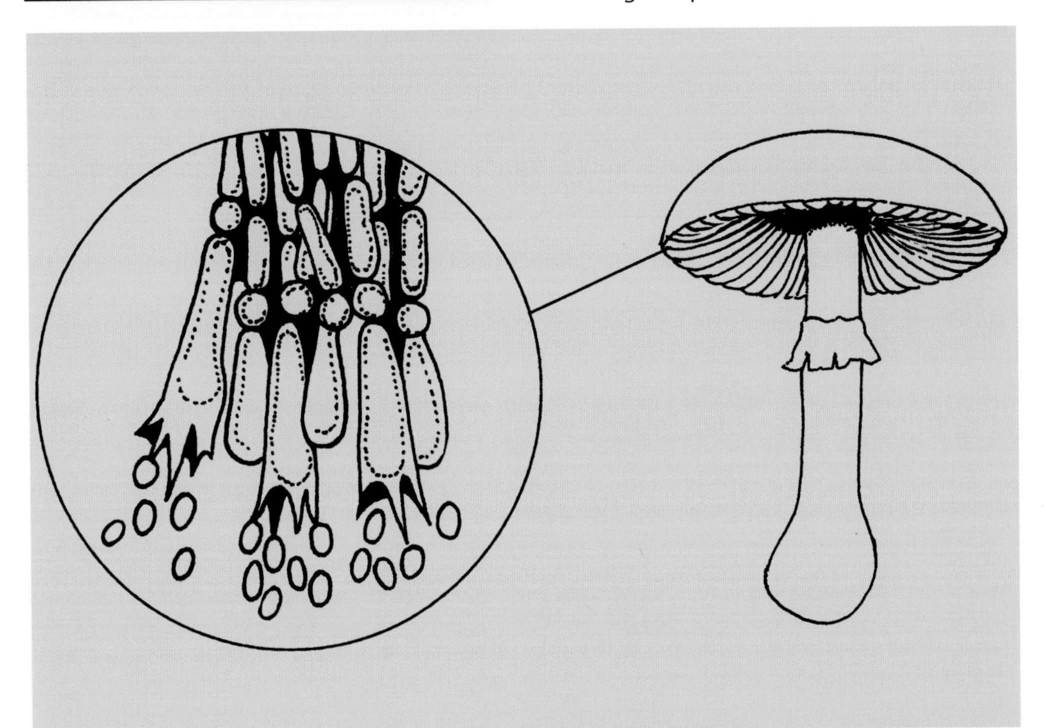

The fruiting body of the Basidiomycetes (spore-shedding) group of mushrooms, which includes the majority of mushroom species. These use both the wind and also animal disturbance to spread their spores.

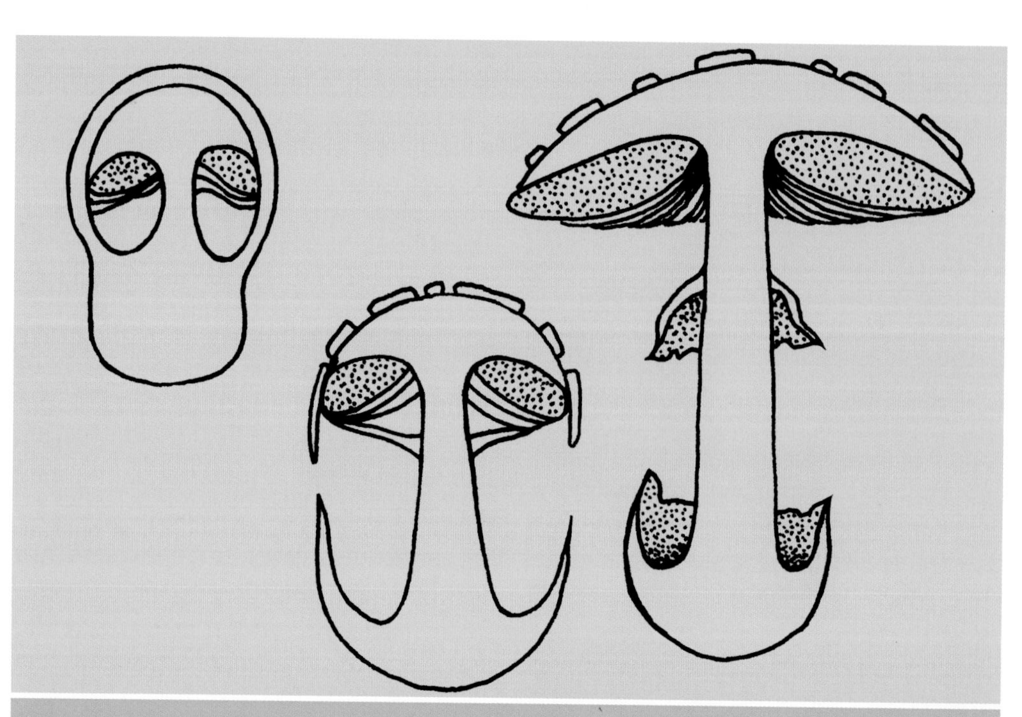

The full volval bag, showing its development from the egg-shaped sac to the finally-emerged fungi, the Amanita family being a good example of this type of fungus. Remnants of the voval bag are left just above the surface of the ground. Note both the ring, halfway up the stem of the mature mushroom, which is the remnant of the lower veil, and also the veil remnants on the cap, which often occur, except in the case of the Death Cap (*Amanita phalloides*).

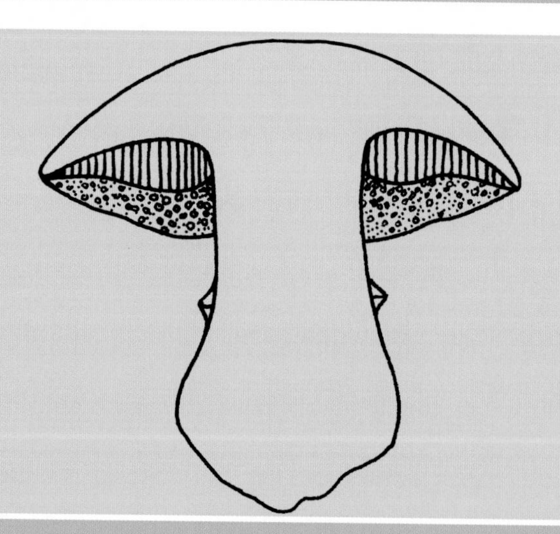

A cross-section of a Boletus-type mushroom, which has pores as opposed to gills. Note the often-enlarged stem, which is equally important from an edibility point-of-view, because it is just as tasty as the cap. The prime example of this kind of mushroom is, of course, the Cep (*Boletus edulis*).

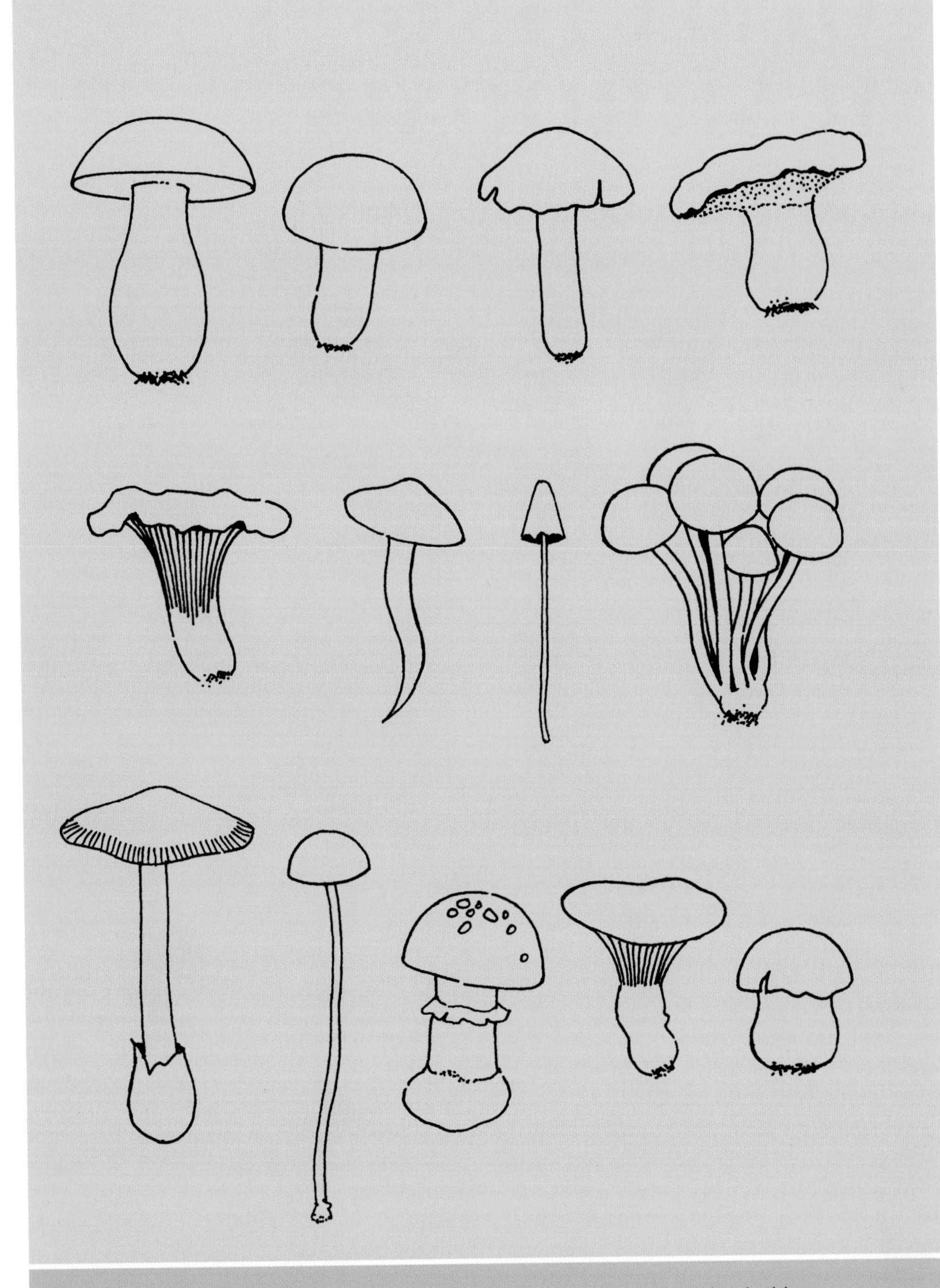

The extremely wide variety of fungal stems, caps, rings and volval bags that you may come across.

COLLECTING MUSHROOMS

Cleaning a catch of Ceps (*Boletus edulis*).

You don't need much equipment for mushroom hunting, but you must go with the right clothing. There is always the weather to contend with, and you are often a long way from shelter.

You need good weatherproof outdoor clothes, with pockets suitable for your pocket guide as well as your mushroom knife and your compass. Good, stout waterproof boots are an essential, and choose good socks, too, for comfort. I prefer breeches to trousers: they are good for walking, and provide protection from brambles and other hazards. It is also important to protect your legs because of the danger of ticks, which can sometimes carry Lyme disease – a viral infection that affects the joints, the heart and the digestive system. I always wear a broad-brimmed hat to protect my eyes and make the mushrooms easier to see: remember the sun is low in the sky in the autumn and therefore often in your eyes.

A good open-weave basket is essential. A basket shaped like a trug is good, as it means that your catch doesn't get crushed. I always have a second basket to put in any mushrooms that need further identification – this is usually carried by my wife, Valerie, when we are on our forays.

On forays I always carry a plastic glove so I can remove a Death Cap (*Amanita phalloides*) (see page 118) or Destroying Angel (*Amanita virosa*) (see page 120)

For successful mushroom hunting:

- Be sure that you have an identification guide, though it is always better to learn from an expert guide on a foray, if you are inexperienced.
- Use a mushroom knife to cut your catch – the most popular ones have a curved blade blunt at the tip as well as a brush. But be careful of legislation – kitchen knives, for example, are not a good idea.
- Always brush the caps to remove loose grit.
- Clean your knife in case of contamination, using disposable wipes.
- Use your thumb-stick to clear away undergrowth.
- Only choose fresh mature specimens to collect.
- Scatter any older specimens to distribute their spores.
- After cutting and brushing, put each mushroom in your basket, cap upwards to stop dirt falling into the gills or pores. This also lets the spores fall through the open weave.

to clearly show the volval bag to the people with me. Should you ever need to handle these mushrooms, always remember to wash your hands straight afterwards, especially before handling other mushrooms (or your sandwiches).

I always carry a thumb-stick to turn over any doubtful mushrooms and to protect against snakes if necessary. Usually snakes will be long gone before you reach them, as they are sensitive to vibration. Sometimes, however, they may be fast asleep, so be careful not to disturb them.

There can be an even more important use for the stick. When I was in Tuscany with Valentina Harris, she told me that once, out with her children hunting mushrooms, she had encountered a wild boar. The advice is never to turn and run when you meet one, as they will give chase if you do. But wild boars do have poor eyesight. Valentina stood her ground and raised the stick to her shoulder so that, to the short-sighted animal, it looked like a rifle. The boar retired quickly. Fortunately, I have never had to use this advice (or my stick for this purpose), although Valerie and I have encountered wild boar more than once on our hunts on the continent.

I always like to carry a good mushroom knife, especially one with a brush. I always like to cut my mushrooms unless it is necessary to remove them totally for perfect identification, and you should always clean mushrooms well before placing them cap up and gills down in your basket. Please be careful as to the type of knife you take on a hunt, in view of current and changing legislation – do ensure that you check this out first.

Happy hunting! And always remember:

IF IN DOUBT, LEAVE IT OUT.

Always follow the Mushroom Pickers' Code:

- On private land, always get permission from the owner.
- Respect the countryside and its crops and animals.
- Leave no litter.
- Avoid damage to vegetation, leaf mould and soil.
- Never remove rare species.
- Wild mushrooms are special: treat them and their habitats with care, so that future generations can enjoy them too.
- If you get a chance, try joining a foray led by an expert: you'll end the day with a greater level of knowledge and experience.

WHEN AND WHERE TO HUNT MUSHROOMS

It is often assumed that mushrooms only grow in the autumn, but this is far from the truth. Good edible mushrooms can be gathered all year. The springtime yields the wonderful St George's Mushroom (*Calocybe gambosa*) (see page 48) and the equally important family of Morels. These are shortly followed by Fairy Ring Champignons (*Marasimus oreades*) (see page 85), and the early summer usually brings a flush of Horse and Field Mushrooms (*Agaricus arvensis* and *Agaricus campestris*) (see pages 28 and 30).

In this book, we have supplied information on the time of year a species occurs in terms of seasons as well as months, for this will depend on your hunting locations. I have also found that it is a good idea to seek local guidance about the mushroom seasons. Local hunters will of course be reluctant to disclose their best locations: they have a secretive nature!

Good Chanterelle (*Catherelis cibarus*) hunting ground.

A St. George's Mushroom (*Calocybe gambosa*) in Spring.

Because mushrooms grow during the hours of darkness, they can sometimes appear as if by magic in the morning. Their amazing growth rate, given the right climatic conditions, can produce a fruiting body of almost a kilogram in just three nights' growth, and a field can often be turned white with Field Mushrooms overnight.

After studying their growth, I am convinced that some of the larger species rely heavily on the phases of the moon for their formidable growth. The gravitational force of the moon allows a quicker uptake of water to the fruiting body, and a large percentage of a mushroom's weight is water, particularly in the larger species. Bearing this in mind, I always have a moon chart in my office. Night-time temperatures and humidity also have a considerable bearing on the growth of many mushrooms (except Bracket mushrooms).

A series of dry years when the seasons are disappointing can be followed by a wetter year and a bumper crop of mushrooms. This happened in 2004, almost all over the UK, resulting in my best find of Ceps (*Boletus edulis*) (see page 44) ever – 160 specimens under two trees in August.

I think it is important to keep a diary of your finds as well as their location, and remember – when you have found a good patch, keep the information to yourself! In this book, we have listed likely habitats to help you in the search, but do remember that this is only a guide. Always keep your eyes peeled as you are foraying.

A November basket.

Where you hunt very much depends on the species you are hunting. For example, some mushrooms are associated with particular species of tree. Some of the bracket mushrooms are almost always found on the same tree type, but they do on occasions find another host. Only 10 per cent of mushrooms grow in open ground, but remember that tree roots can often travel a long way.

You will quickly gather experience from your own observations. Add to it the help of a foray with an expert, and you will soon be a very successful hunter.

MUSHROOM FORAYS

In the previous chapters, you may have noticed several references to mushroom forays. A foray is an excellent way to increase not only your confidence but also your mycological knowledge. Interest in foraying has increased dramatically in the last few years – one reason for this being a growth of interest in the countryside, another being a desire to use leisure time on stimulating pursuits that include healthy exercise.

There are different kinds of foray, each with their own particular focus: some, for example, concentrate on identification, while others are more interested in finding good mushrooms for the pot.

Valerie and I have been conducting forays for over 15 years. In that time we have taken hundreds of people foraying, both in Britain and abroad. We have learned from experience that an ideal group contains about 15 people: a good number for mixing well – and for a united team of mushroom spotters. The more eyes you have, the more you find.

We took a Swedish eye surgeon out on a foray a few years ago, and he assured me that foraying was very good for the eyesight.

'You're continually using your peripheral vision,' he explained, 'and you're strengthening your eye muscles as well.'

Yet another good reason for mushroom hunting then!

Forays also provide something else of enormous value: they offer you a health-giving and relaxing break from the hectic nature of modern life. You also get to see marvels you might otherwise miss. On one of my forays recently we saw two large red deer fighting during the rut, and in the evening of the same memorable day we watched a family of badgers with their young on a hunting exercise. Sightings of rare birds can be frequent, so it's also worth having a pair of binoculars in your bag as well as a camera.

I have stressed the need to take two baskets with you when you are foraying. One basket is for the mushrooms you have identified and know are edible. The second is for specimens that need further checks for certain identification. It really is essential to keep the clearly recognised mushrooms separate from those you're not sure of, and also essential to have both of your baskets checked by the foray leader before you go home to feast on your finds.

There has been a lot of concern in recent years about over-exploitation of mushrooms in the wild, but my own observations tend to suggest that, far from diminishing the number, correct collection with baskets has effectively spread spores and increased mushroom crops. We have been taking people on forays in one particular East Anglian wood for over 15 years. Because of the nature of this woodland, we usually take a similar route through it, and have noticed that there are now far more mushrooms than there were 15 years ago. An experiment by the British Mycological Society, conducted over five years with two control plots, also confirmed that the numbers of mushrooms have grown. They gathered all of the fungus caps on one plot but took none from the second, and at the end of the five years there was a significant increase in the number of fruiting bodies in the plot that had been regularly harvested.

Many of us have Internet access these days. You can find numerous references to mushroom forays on the web, and even discover holidays with a focus on mushroom hunting and cooking. We have been involved with some of these, not only in Britain but also in France and Italy. Courses involving both hunting and cooking with mushrooms are great fun: for a start, you get the chance to increase your cooking skills, while enjoying the company of others of a like mind. Then there are visits to French and Italian markets and vineyards, which are not to be missed. The array of wild mushrooms on sale in the French and Italian markets is a delight to see. Closer to home, pay a visit to London's Borough Market – there you'll find a splendid variety of wild mushrooms, both fresh and dried.

I hope you now feel inspired to go on a foray yourself. There are many recording groups and associations connected with foraying and we have listed some of these for you at the back of this book. And don't forget – there is always the catch to bring home and turn into a delicious meal!

On the fungi trail.

HOW TO USE THIS BOOK

Boletus edulis
Cep or Penny Bun (see page 44)

Agaricus campestris
Field Mushroom (see page 30)

EXCELLENT VERY GOOD GOOD FAIR

Agaricus arvensis
Horse Mushroom (see page 28)

Polyporus squamosus
Dryads Saddle (see page 94)

In this book, we have set out to give you a clear and concise list of the species you would like to find for the pot, along with possible lookalike species in order to help you avoid mistakes. The poisonous species covered are those that we felt you should recognise for safe collecting. We have used a colour code system for easy reference – green for Edible and red for Poisonous.

There are thousands of fungi occuring in the wild, and so this is by no means a definitive guide. In the *Further Reading* section (see page 152), we provide a list of titles that you may like to have at home for further study.

Each of the featured species has an 'Edibility rating' – from 'Excellent' through to 'Deadly Poisonous' – along with typical season, habitat, size and main identifica-

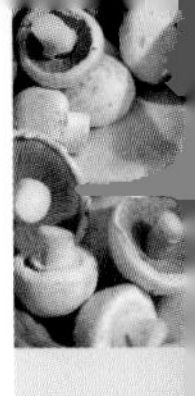

Boletus erythropus
(see page 123)

Amanita phalloides
Death Cap (see page 118)

DANGEROUS — POISONOUS — DEADLY POISONOUS

Lactarius pubescens
Var. Woolly Milk Cap (see page 134)

Amanita virosa
Destroying Angel
(see page 120)

tion features. For additional guidance, the text is illustrated with clear photographs. All of these images were shot in available light, in order to depict the conditions that mushroom hunters may experience in the field as closely as possible. The aim was also to provide attractive images showing specimens in their natural habitat: in identifying mushrooms, where they grow can be every bit as important as their appearance.

Finally, to help you with the culinary side of things, we have also provided some advice on the best ways of cooking and storing your finds. These are covered in greater depth in the *Storage Methods* and *Cooking With Mushrooms* sections (see pages 144 and 147).

We hope you enjoy using this guide to further your enjoyment of sustainable wild mushroom hunting.

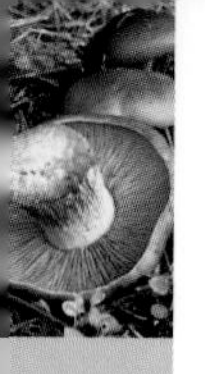

EDIBLE SPECIES

A pair of Ceps (*Boletus edulis*).

This section deals with what I consider the best of the edible species.

If you look in encyclopaedic mushroom guides, you'll find plenty of mushrooms listed as 'edible', but in my experience most of them are simply not worth collecting.

I have been cooking with wild mushrooms for over 40 years now, and have had good results with all the species I include here. They were a major feature on the menu when we ran our pub restaurant in Burnham Market.

I have listed the edibility in grades from 'Excellent' down to 'Fair', to give you an idea of their importance in the kitchen. (For more information on the grades, see *How to Use This Book*, page 24.)

Never forget that mushrooms are a rich food: even edible ones can sometimes cause stomach upsets. So take care if you are serving mushroom dishes to guests.

I have provided a detailed description of each mushroom, as well as information about its season, habitat and size. There is also an at-a-glance summary of its main identification features, along with suggestions about how it can be cooked (or stored for use later).

I have taken care to give details of any lookalikes which you should avoid. There is more information about these in the *Poisonous Species* section (see page 112).

The more you learn about wild mushrooms, the more careful you will become. The one way to pick them with safety is to identify them correctly and with certainty.

Only collect mature specimens in order to avoid mistaken identities – when still young, some varieties are not always easy to tell apart. Choosing mature specimens is environmentally friendly, too: they will be shedding spores, which will remain to become the next season's crop.

Always carry two baskets – one for the mushrooms you have identified and one for any you want to check later. That way there's no risk of contamination.

One of the best ways to increase your knowledge is to join a mushroom foray led by an expert. There are details about this in the *Mushroom Forays* chapter (see page 22). Enjoy your mushroom hunting and don't take any risks with mushrooms you cannot identify.

IF IN DOUBT, LEAVE IT OUT.

Agaricus arvensis
Horse Mushroom

EDIBILITY Very good

SEASON CHECK
Mid-June to late October
(Early summer to autumn)

HABITAT
Mature grazing pasture with chalky soil.

SIZE
Cap 8–20cm across. Stem 8–10cm high, 2–3cm thick.

MAIN IDENTIFICATION FEATURES
Often grows in rings, which are broken. Pleasant aniseed smell. Sometimes very large with a ring on the stem.

This large mushroom is meaty in texture and also has a distinctive aniseed smell. If the creamy white cap has yellow markings, check that it's not the poisonous Yellow Stainer (*Agaricus xanthodermus*) (see page 115), which bruises bright yellow when cut or scratched (this is very noticeable on the stem). However, do bear in mind that the Horse Mushroom can have a yellow tinge to the cap and stem. The young mushrooms have white gills, growing to delicate pink and, later, darker brown. The flesh is thick and white. Older caps can turn brassy yellow with age or when wet. Pick them young: older ones are often maggoty and the flesh turns dark brown, staining anything they are cooked with.

Cooking suggestions
The young ones are delicious simply sliced and fried. Older ones make a fine addition to savoury dishes, and add their own dark brown colouring. You can use them to make your own thick, homemade ketchup.

Best storage method
Dried.

WARNING

Beware of the Yellow Stainer (*Agaricus xanthoderma*) (see page 115), which often grows with and at the same time as the Horse Mushroom. The base of the Yellow Stainer's stalk turns bright yellow when cut or scratched and it has an unpleasant smell of ink.

Agaricus augustus
The Prince

EDIBILITY Excellent

SEASON CHECK
Mid-June to late October
(Early summer to autumn)

HABITAT
Along the edges of coniferous and deciduous woodland. Often along hedges, especially if Holm Oak grows nearby.

SIZE
Cap 10–20cm across. Stem 10–20cm high, 2–4cm thick.

MAIN IDENTIFICATION FEATURES
Scaly brown cap and large size. Distinctive smell – similar to almonds. Often grows in groups rather than rings.

This fine mushroom is rightly called 'the Prince' for its fine flavour. It is quite large and tends to grow in groups. The cap is yellowish-brown and covered with chestnut brown scales. There is a distinct ring below the cap. The gills are white when young, turning to darker brown with age. The flesh is thick and white and develops tinges of red with age. The smell is distinctive and quite strong, very like bitter almonds. As with all Agaricus, it is best picked young, as the older specimens are often maggoty.

Cooking suggestions
These mushrooms are delicious fried and also added to stews. They go well with pasta dishes, and are great for mushroom risotto.

Best storage method
Dried.

Agaricus campestris
Field Mushroom

EDIBILITY Good

SEASON CHECK
July to September
(Summer to early autumn)

HABITAT
Open fields and headland with short grass, often along cliff edges. Playing fields and village greens can be productive too.

SIZE
Cap 3–10cm across. Stem 3–10cm high, 1–2cm thick.

MAIN IDENTIFICATION FEATURES
Starts with pink gills, which darken with age to chocolate brown or almost black. Grows in clusters and in rings in short grass. Fragrant smell.

This is probably the best-known wild mushroom, less widespread now that farming practices have cut down its habitat. Best picked in the early morning and when the caps are still dome-shaped. The caps are silky white to cream and can be scaly or smooth; they age to light brown. The young gills are deep pink, later maturing to brown. The flesh bruises slightly pink and smells pleasantly mushroomy. The white stem sometimes yellows slightly towards the tapered base. These mushrooms can be dried and stored without losing their flavour.

Cooking suggestions
No need to peel them – just wipe, slice and fry! Good in any mushroom dish and almost unbeatable with bacon and eggs for breakfast. Try them raw in salads too. Older specimens make a gloriously deep brown ketchup or sauce.

Best storage method
Dried (slice the larger ones).

Agaricus silvicola
Wood Mushroom

EDIBILITY Good

SEASON CHECK
September to November
(Autumn)

HABITAT
Mainly coniferous and deciduous woodland, but also on woodland edges and along paths close to the wood.

SIZE
Cap 5–10cm across. Stem 5–8cm high, 1–1.5cm thick.

MAIN IDENTIFICATION FEATURES
White cap, which turns yellow with age. Large ring and a characteristic smell of aniseed.

As the name suggests, this mushroom grows in the woods. The cap is almost flat. It is white when young and yellows with age. The gills are pale pink at first, turning brown with age. There is a large ring that stays on the stem after the veil has broken. The smell is of aniseed. Be careful not to confuse with the Yellow Stainer (*Amanita xanthodermus*) (see page 115).

Cooking suggestions
With its slightly different taste, it is very good in mixed mushroom dishes and especially with pasta.

Best storage method
Dried after slicing.

WARNING

Beware of the Yellow Stainer (*Agaricus xanthoderma*) (see page 115), which often grows in the same location as the Wood Mushroom. The base of the Yellow Stainer's stalk turns bright yellow when cut or scratched and it has an unpleasant smell of ink.

Agaricus urinascens (A. macrosporus)

EDIBILITY Good

SEASON CHECK
June to October
(Early summer to autumn)

HABITAT
Usually pasture, growing in rings (but can also grow in clusters).

SIZE
Cap 8–25cm across. Stem 5–10cm high, 2–3cm thick.

MAIN IDENTIFICATION FEATURES
Very domed cap with whitish gill at first. Smells of ainseed.

This mushroom is similar to the Field Mushroom (*Agaricus campestris*) (see page 30) in many respects, except that it is larger and thicker. The cap is dome-shaped and white, turning a little brown with age. The gills start whitish grey, colouring dark brown with age. There is a faint smell of aniseed, which becomes more like ammonia with age. It grows in rings and has a distinct veil when young, which breaks to reveal the light-coloured gills.

Cooking suggestions
With its firm texture, it is good in all mushroom dishes. Young ones are also good served sliced raw in salads.

Best storage method
Dried after slicing.

Aleuria aurantia
Orange Peel Fungus

EDIBILITY Good

SEASON CHECK
September to November (Autumn)

HABITAT
From bare earth to short grass along woodland edges, growing in quite large clumps.

SIZE
Cap 0.5–5cm across.

MAIN IDENTIFICATION FEATURES
Distinctive shape, with wavy, flower-like edges and a very insignificant stem. Vivid orange colour. No characteristic smell.

This wonderfully bright-coloured fungus has a nice taste and texture, and gives an exciting look to your wild mushroom dishes. The cap is cup-shaped with wavy edges and is bright orange in colour. It grows in clusters in grassland and along roadside edges in woods. As the name suggests, it looks very like discarded orange peel.

Cooking suggestions
This makes a wonderful addition to any mixed mushroom dish, not only for its colour but also for its interesting texture.

Best storage method
Dried.

Amanita caesarea
Caesar's Mushroom

EDIBILITY Excellent

SEASON CHECK
July to October
(Summer to autumn, especially after summer storms)

HABITAT
Open deciduous woodland, especially with oak or Sweet Chestnut. Usually in southern Mediterranean regions.

SIZE
Cap 6–18cm across. Stem 5–12cm high, 1.5–2.5cm thick.

MAIN IDENTIFICATION FEATURES
Orange-to-yellow cap with the distinct volval sac and yellow stem. Grows in small groups.

This is an Amanita mushroom highly prized for its excellent flavour, but great care must be taken in collecting to avoid confusion with poisonous members of this dangerous family. This mushroom has not been found in the UK as yet, but with global warming, it could happen soon.

It hatches out of a volval egg sac – hence its Italian name, *ovolo*. It is often sold in markets still in this egg form, which means that the poor mushroom has had no time to shed its spores before being picked. As a result there has been a rapid decline in numbers.

The volval sac splits to reveal a deep red cap; the cap's colour changes as the mushroom grows, from orangey-red to a light orange when mature. There are often traces of the volval bag on the cap. The cap is hemispherical to convex and is finely lined at the margin. The stem is light yellow with a large yellow ring. The gills are yellow and crowded. The flesh is whitish and the aroma pleasant.

Cooking suggestions
This is wonderful in all mushroom dishes but can be eaten raw to get the full flavour. There are many recipes for its use, as it has been popular since ancient Roman times.

Best storage method
Dried.

Fly Agaric (Amanita muscaria).

> **WARNING**
>
> **Particular attention must be taken to avoid confusion with a juvenile Fly Agaric (*Amanita muscaria*), which has similar colours and grows in similar places at the same time (see page 117).**

Amanita fulva
Tawny Grisette

EDIBILITY Good

SEASON CHECK
August to November
(Late summer to late autumn)

HABITAT
Mixed woodland.

SIZE
Cap 4–9cm across. Stem 7–12cm high, 0.75–1.2cm thick.

MAIN IDENTIFICATION FEATURES
Delicate orange-brown cap with distinct grooves on the edge. Narrow volval bag and hollow stem.

This is another member of the Amanita family worth including in your mushroom basket. Remember that some Amanitas are dangerous, including the Death Cap (*Amanita phalloides*) (see page 118).

This Amanita, however, is edible and delicate, and tends to grow singly in woodland habitats. The cap is a pale orange-brown with distinct grooved edges. The stem is slender and hollow, without markings. The gills are white and once the veil breaks there is no distinct ring. The narrow volval bag is easy to see and so helps in identifying this species.

Cooking suggestions
This is a delicate mushroom so it is often best used on its own. I have got the best results in omelettes. You can also add it to mixed wild mushroom dishes.

Best storage method
Dried.

Amanita rubescens
The Blusher

EDIBILITY Fair

SEASON CHECK
June to November
(Early summer to late autumn)

HABITAT
Both deciduous and coniferous woodland, where it is very common.

SIZE
Cap 5–15cm across. Stem 6–14cm high, 1–2.5cm thick.

MAIN IDENTIFICATION FEATURES
Flushed, reddish-brown cap with bulbous stem, which blushes when bruised.

This is another of the dangerous Amanita family, but it is edible and very common. Great care must be taken in identification for possible confusion with poisonous Amanitas.

The Blusher grows out of a bulbous base, which terminates in a ridge or gutter. The cap is a rosy brown or flesh colour, sometimes with yellowish flush and white or reddish patches. The thick stem is white but, below the ring, it flushes red. The flesh is white, but slowly blushes pink when bruised or exposed to air. The gills are white, and they too turn red when bruised. The aroma is pleasant and the flavour is mild.

Cooking suggestions
These mushrooms are toxic if eaten raw or partly cooked, so you should always blanch them in boiling water and discard the water before final cooking. As the mushroom has a mild taste it is best used in mixed mushroom dishes.

Best storage method
Dried after blanching.

WARNING

Be careful not to confuse this mushroom with the (possibly deadly) poisonous Panther Cap (*Amanita patherina*) (see page 122). Also, note that this mushroom should never be eaten raw or partly cooked.

Armillaria mellea
Honey Fungus

EDIBILITY Good

SEASON CHECK
July to November
(Summer to late autumn)

HABITAT
Often older woodland, especially near oak and beech. Also likes tree stumps.

SIZE
Cap 3–15cm across. Stem 6–15cm high, 0.5–1.5cm thick.

MAIN IDENTIFICATION FEATURES
Grows in large clusters on both live and dead trees, and tree stumps – particularly oak.

This nightmare for gardeners is sometimes called Bootlace Fungus: it grows from black cords found under tree bark, which travel great distances underground to find new trees to colonise. Honey Fungus kills off woodland and shrubberies, but its caps (when cooked) are edible. It varies greatly in appearance. Caps start off convex, then flatten and depress to dish-shaped. The gills range from off-white to dark brown. The flesh (which is honey-coloured to deep brown) smells strong and sweet.

Cooking Suggestions
Use only young caps for eating (even the young stalks are tough). Be sure to blanch them first in lightly salted boiling water – and throw away the water afterwards. They are very good sautéed in butter with onion, garlic, fresh basil and cream and served with pasta.

Best storage method
Does not dry well, so freeze made-up dishes.

WARNING

Beware of confusion with Sulphur Tuft (*Hypholoma fasciculare*) (see page 132). Also, keep these mushrooms in an airtight container, when bringing them home, to prevent spreading spores in your garden.

Auricularia auricula-Judea
Jew's Ear

EDIBILITY Fair

SEASON CHECK
Virtually all year round

HABITAT
Almost exclusively on dead or dying elder trees. Also found on beech.

SIZE
Fruit body 2–8cm across.

MAIN IDENTIFICATION FEATURES
Ear-shaped fruit body. Hard in dry weather but otherwise rubbery.

Sometimes called Tree Ear, this mushroom was named after Judas, who is said to have hanged himself from an elder tree, where the species is frequently found throughout the year.

Young mushrooms are pale brown, but darken with age to a rich tan or even a purplish brown. They have a velvety outer surface and a jelly-like texture, which hardens and wrinkles with age, feeling elastic and rubbery between the fingers. Tiny greyish hairs on the smooth inner surface emphasise the mature mushroom's likeness to an elderly human earlobe.

Cooking suggestions
Wash really well, slice thinly and cook thoroughly to ensure that the mushrooms are tender. Try them in a thick garlic-and-basil-flavoured cream sauce for filling vol-au-vents. They are good in stews too.

Best storage method
Dried. In dry weather they can be collected dry and stored in airtight containers.

Boletus aereus

EDIBILITY Excellent

SEASON CHECK
August to October
(Late summer to autumn)

HABITAT
Near broad-leaved trees, especially oak and beech.

SIZE
Cap 7–16cm across. Stem 6–8cm high, 4–6cm thick at the widest point.

MAIN IDENTIFICATION FEATURES
Large and bulbous brown cap with a shiny surface. Wide brown stem with a network of markings.

One of the best of the Boletus family to find. Although quite rare, in a good year it can be found in large quantities. The dark brown cap is markedly dome-shaped. The stem is very bulbous and quite short with a network of brown markings. The flesh is white. The tubes are white to cream, turning eventually dull sulphur yellow. The smell is earthy and strong.

Cooking suggestions
A highly versatile mushroom with great flavour and texture, which will enhance any dish. Good raw, too, thinly sliced in a salad with a warm walnut oil dressing and a few shavings of parmesan on top.

Best storage method
Sliced thinly and dried.

Boletus appendiculatus

EDIBILITY Excellent

SEASON CHECK
August to October
(Late summer to autumn)

HABITAT
Near broad-leaved trees, especially oak. More common in southern England and on the continent.

SIZE
Cap 8–14cm across. Stem 11–12cm high, 3–4cm thick.

MAIN IDENTIFICATION FEATURES
Rust brown cap with an unusual shape. Yellow stem that bruises blue when cut.

Another great Boletus for your basket. Unlike the quite rare *Boletus aereus* (see page 39), this is one to look out for. The large cap is rust-brown in colour and can have irregular cracks near the centre. Its shape is often irregular, sometimes almost heart-shaped. The stem is bright yellow, sometimes with red patches. The flesh is white to pale yellow, bruising blue when cut. The tubes are lemon yellow. The smell is pleasant, rather like that of Puff-balls.

Cooking suggestions
This is a very good Boletus and young, firm specimens can be sliced and eaten raw. It is very good in all your wild mushroom dishes.

Best storage method
Sliced and dried.

Boletus chrysenteron
Red Cracked Boletus

EDIBILITY Fair

SEASON CHECK
August to November
(Late summer to late autumn)

HABITAT
Mixed woodland and along tree-lined drives and paths. Likes beech.

SIZE
Cap 4–11cm across. Stem 4–8cm high, 1–1.5cm thick.

MAIN IDENTIFICATION FEATURES
Cracked cap with pinkish tinge. Lemon yellow pores.

The caps, velvety when young but growing smoother, range from buff to dull brown, sometimes taking on a reddish flush in cold weather. They often have cracks showing pinkish red flesh underneath. Another identification feature is the stem, which is tinged distinctly red for most of its length (although there are some other Boletuses with this feature). The pores are yellow and open, and stain light green when bruised. The flesh is cream to yellow, does not bruise when cut, and is less dense in texture than that of the Bay Boletus (*Boletus badius*) (see page 42) or the Cep (*Boletus edulis*) (see page 44).

Cooking suggestions
These are less dense than many Boletuses and are best used from dried, as this intensifies their flavour. A good addition to soups and stews.

Best storage method
Dried.

Boletus badius
Bay Boletus

EDIBILITY Excellent

SEASON CHECK
August to November
(Late summer to late autumn)

HABITAT
Mixed woodland, especially where there is moss.

SIZE
Cap 4–14cm across. Stem 4.5–12.5cm high, 0.8–4cm thick.

MAIN IDENTIFICATION FEATURES
Dark velvety cap. Pores, not gills, which stain blue-green when bruised.

Bay Boletus looks felty when young, becoming smooth and polished with age. These mushrooms are fairly viscid when damp. Cap and stem are both dark, ranging from shades of bay to deep brick brown. The pores are light yellow, staining blue-green almost immediately when pressed or cut – this is a feature of other Boletuses too. The white flesh has a mild mushroomy smell and stains bluish when cut, but the stain fades. Excellent flavour can be obtained from clean specimens picked when dry. They are usually maggot-free!

Cooking suggestions
A really versatile mushroom: it dries very well and tastes delcious. Try them raw, thinly sliced. They go well with game dishes, such as pheasant and partridge.

Best storage method
Very good dried but can also be stored under oil or salt.

When wet, Bay Boletus can resemble Slippery Jacks (*Suillus luteus*) (see page 102).

Boletus edulis
Cep or Penny Bun

EDIBILITY Excellent

SEASON CHECK
Late July to late November
(Summer to late autumn)

HABITAT
Coniferous, broad-leaved and mixed woodland, often by pathways. Likes dappled shade and can often be found hiding under ferns.

SIZE
Cap 8–30cm across. Stem 3–23cm high, 3–7cm thick.

MAIN IDENTIFICATION FEATURES
Large light brown cap with bulbous stem. White pores turning greyish yellow with age. Dense and solid feel.

A mushroom prized for its excellent nutty flavour. The young mushroom's light brown cap suggests freshly-baked bread – hence the name. The caps darken with age and their velvety sheen looks sticky in wet weather. The very bulbous stem has a fine network of white markings, most clearly seen closest to the cap. The pores are white, later turning light yellow. The flesh is always white. Ceps are best collected for eating when small and tight, although they can grow very big – their caps developing a characteristic white rim in the process. I have collected specimens over 1kg in weight and recently collected 160 beneath a single tree – the find of a lifetime. Large specimens can often be attacked by flies and hence become maggoty, so it is best to check for holes before adding to your basket.

Cooking suggestions
Possibilities with the Cep are endless, as it is one of the most widely used mushrooms today. Clean well by wiping and cut to check that they are perfect. They are best when small and tight. Remember that the stem has as much flavour as the cap. Great eaten raw in salads, and a perfect addition to all pasta dishes, to which parmesan cheese adds the final touch.

Best storage method
Dried. (Dried Ceps can often be found in markets and stores throughout the world – a very good buy.)

Boletus pruinatus

EDIBILITY Good

SEASON CHECK
September to November
(Autumn)

HABITAT
Under broad-leaved trees in dappled shade, especially beech. Also likes mossy banks.

SIZE
Cap 4–10cm across. Stem 4–8cm high, 2–3cm thick.

MAIN IDENTIFICATION FEATURES
Dark brown velvety cap with bright yellow pores, which bruise slowly bluish.

This Boletus is a good find: it has a lovely colour and is quite dense, making it excellent in mixed mushroom dishes. The cap is a dark velvety brown, becoming lighter with age. The stem is lemon yellow with a scattering of tiny red dots. The flesh is yellow, darker near the base of the stem, and slowly turns a bluish green when cut. The dense pores are lemon yellow, becoming bluish with age.

Cooking suggestions
This is a good addition to all mixed mushroom dishes. It has a soft texture when cooked. Use it dried, as this intensifies the flavour.

Best storage method
Dried.

Bovista plumbea
Puff–ball Var.

EDIBILITY Good

SEASON CHECK
April to November
(Spring to late autumn)

HABITAT
Lawns with short grass and short pasture – especially golf courses.

SIZE
Fruit body 4.5–6cm across.

MAIN IDENTIFICATION FEATURES
Almost perfect round shape with smooth outer skin.

The Puff-ball Var. is one of the more common forms of mushroom you may come across. It should only be eaten when firm, white and young. Always cut it in half to check it is indeed firm and white throughout. (There are several members of the Puff-ball family in this book, under various scientific names, and this is the first.) It is often perfectly round with only a few strands holding it in the soil. Its outer surface is smooth. Puff-balls often grow in large numbers.

Cooking suggestions
Good fried for breakfast with bacon and eggs, or added to any wild mushroom dish.

Best storage method
Dried.

Calocybe gambosa
St George's Mushroom

EDIBILITY Excellent

SEASON CHECK
April to May and sometimes September (Spring and sometimes early autumn)

HABITAT
Chalk grassland, open commons and old airfields.

SIZE
Cap 5–15cm across. Stem 2–4cm high, 1–2.5cm thick.

MAIN IDENTIFICATION FEATURES
Chalky white cap. Distinctive mealy aroma. Often grows in large rings.

This is a spring mushroom and heralds the start of the season. It grows in long-established rings that can be very large. Young caps are well rounded but grow irregular and wavy when larger and older. The cap is coloured white to cream and has an in-rolled rim. The flesh is white, thick and soft with a pleasant mealy taste and aroma. The whitish gills are narrow and crowded and the white stem is tough and fibrous. These mushrooms are particularly good for drying.

Cooking suggestions
The fine flavour of this mushroom can be really appreciated if you eat it raw, perhaps with an olive oil dressing and a dash of garlic or chives. Cooked, the flavour is delightfully nutty and goes well with chicken or fish. Use it to make a mouth-watering chicken casserole.

Best storage method
Dried after slicing.

Calvatia gigantea (Langermannia gigantea)
Giant Puff-ball

EDIBILITY Good

SEASON CHECK
July to October
(Summer to autumn)

HABITAT
Often grazing marshes and open fields, and also along drainage ditches.

SIZE
Fruit body 7–80cm+ across. (Some can even grow up to 1m across.)

MAIN IDENTIFICATION FEATURES
Creamy white cap. Can often be seen from a distance because of its size (it is the largest of the Puff-ball family).

'Gigantea' is a well-earned epithet: Giant Puff-balls can grow to an enormous size. However, they are only edible when young. Fresh specimens sound hollow when you tap their caps and their flesh is solid and white – a knife should be able to cut them cleanly without tearing. Do not pick them if the flesh is discoloured or if the outer wall has broken down to reveal an exposed yellow mass of spores. Eventually the fruit-globe will break free and roll around, spreading the spores far and wide.

Cooking suggestions
Only use Puff-balls with a pure white centre. These young ones are excellent in all wild mushroom dishes, soups and stews. They are a fine breakfast mushroom too: slice and fry with bacon, or dip in beaten egg and breadcrumbs and fry lightly. They will keep fresh for several days, cling-filmed and stored in the refrigerator.

Best storage method
Dried after slicing into cubes. Made-up dishes also freeze well.

Cantharellus cibarius
Chanterelle

EDIBILITY Excellent

SEASON CHECK
July to November
(Summer to late autumn)

HABITAT
Coniferous and deciduous woodland, often on mossy banks.

SIZE
Cap 3–10cm+ across. Stem 3–8cm high, 0.5–1.5cm thick.

MAIN IDENTIFICATION FEATURES
Striking yellow cap with fluted shape and apricot aroma.

True Chanterelles, solitary or in clusters, are a great find for both their appearance and their superb flavour. Flat at first with a broken in-curved rim, the cap later becomes concave and fluted, like a single trumpet-shaped flower. The solid stem tapers at the base. The yellow blunt-forked ridges are narrow and irregular. The flesh ranges from pale to deep egg yellow, but can fade with age; sometimes it turns quite orange. There is a faint aroma of apricots, which is often difficult to detect.

WARNING

Beware of the False Chanterelle (*Hygrophoropsis aurantiaca*) (see page 131), which has gills instead of blunt-forked ridges, and also Red Staining Inocybe (*Inocybe erubescens*) (see page 133).

Cooking suggestions

True Chanterelles are wonderful tossed in caramelised butter, but their flavour and texture survive slow cooking too. Trim the bases and then cook them whole. They are endlessly versatile: try them with meat and fish, use them in mixed mushroom dishes, or add to sauces (they provide a marvellous colour as well as taste).

Best storage method

Good dried, but also try storing in spiced alcohol, extra virgin olive oil or vinegar.

Cantharellus infundibuliformis
Winter Chanterelle or Yellow Leg

EDIBILITY Excellent

SEASON CHECK
September to November
(Autumn)

HABITAT
Mixed woodland with fallen branches, under leaf litter, often where there is moss, and especially near Sweet Chestnut trees.

SIZE
Cap 2–5cm across. Stem 5–8cm high, 0.4–0.9cm thick.

MAIN IDENTIFICATION FEATURES
Brown cap that is quite small and fluted with a hollow yellow stem (which gives the Winter Chanterelle its alternative name, Yellow Legs).

So called because they appear much later than ordinary Chanterelles, these mushrooms grow in clusters, often under fallen leaves, blending in with them, and so making the mushrooms hard to spot. The dark brown caps are convex at first, then become funnel-shaped with a fluted rim. The yellow stem quickly becomes hollow. The blunt-forked ridges are narrow and irregularly branched; yellow at first, they age to greyish blue. The thin yellowish flesh has a sweet aroma of honey and a sweet flavour when cooked. Cut the stalks with a knife or scissors to keep the caps clean.

Cooking suggestions
Another versatile mushroom, perhaps at its best in mixed mushroom dishes. Trim the bases and cook whole. The pleasingly sweet flavour goes well with fish.

Best storage method
Dried, but can be stored in extra virgin olive oil or white wine vinegar.

> **WARNING**
>
> **Beware of the False Chanterelle (*Hygrophoropsis aurantiaca*) (see page 131), which has gills instead of blunt-forked ridges, and also Red Staining Inocybe (*Inocybe erubescens*) (see page 133).**

As with Chanterelles (*Cantharellus cibarius*) (see page 50), Winter Chanterelles can be almost impossible to see amongst leaf litter.

Chalciporus piperatus
Peppery Boletus

EDIBILITY Good

SEASON CHECK
August to October
(Late summer to autumn)

HABITAT
Usually birch scrub or mixed pine woodland.

SIZE
Cap 3–7cm across. Stem 4–7cm high, 0.5–2cm thick.

MAIN IDENTIFICATION FEATURES
Grows in small groups, often in birch scrub. Reddish pores. When cut, the stem is distinctly yellow towards the base.

This Boletus is collected as a flavouring rather than as a main ingredient for a dish. It is also one of the smallest Boletuses that you can pick. The cap is cinnamon to sienna in colour, and the pores are red. The stem is cinnamon-coloured, showing yellow when cut, and is tapered at the base.

Cooking suggestions
Use only as flavouring, and sparingly – it is quite peppery, as the name suggests. I dry all those that I find and use them as a condiment in various dishes.

Best storage method
Dried.

Clitocybe odora
Aniseed Toadstool

EDIBILITY Good

SEASON CHECK
September to November (Autumn)

HABITAT
In leaf litter along the edges of both deciduous and coniferous woodland. Favours beech and Sweet Chestnut.

SIZE
Cap 3–7cm across. Stem 3–6cm high, 0.5–1cm thick.

MAIN IDENTIFICATION FEATURES
Blue-green cap and the clearly recognisable aroma of aniseed.

The Aniseed Toadstool is most useful as a flavouring, used after drying. The cap is button-shaped at first, but it soon flattens out and sometimes becomes wavy. The colour is blue-green, darkening with age. It also smells strongly of aniseed, which clearly identifies it.

WARNING

Beware of its lookalike: the Verdigris Agaric (*Stropharia aeruginosa*) (see page 142).

Cooking suggestions
The caps can be used fresh – but only sparingly: the flavour of aniseed is intense. This goes well with fish dishes in a sauce. I use the caps dried and powdered in many dishes that require aniseed flavour.

Best storage method
Dried and powdered.

Clitopilus prunulus
The Miller

EDIBILITY Good

SEASON CHECK
August to late October
(Late summer to autumn)

HABITAT
Usually open grassland in woodland areas, growing in troops.

SIZE
Cap 3–12cm across. Stem 1.5–2.5cm high, 0.4–1.2cm thick.

MAIN IDENTIFICATION FEATURES
Creamy white cap and a strong mealy smell.

These may be small but they are very tasty mushrooms. Great care must be taken in collecting to avoid confusion between these and the *Clitocybe dealbata* and *C. rivulosa* (see pages 126 and 127): they are very similar in appearance and grow in similar places.

The cap is creamy white with a greyish tint, and becomes irregular and wavy with age. The gills are white, turning to pink as they mature. The aroma of damp flour is very distinctive and a good identification feature. The smell is what gives this mushroom its name.

Cooking suggestions
A good mushroom to add to all your mixed mushroom dishes.

Best storage method
Dried.

Clitocybe dealbata and C. rivulosa

WARNING

Be very careful not to confuse this with the very similar poisonous species *Clitocybe dealbata* (*above left*) and *C. rivulosa* (*above right*) (see pages 126 and 127).

Coprinus comatus
Shaggy Ink Cap

EDIBILITY Good

SEASON CHECK
July to September
(Summer to early autumn)

HABITAT
Open ground, often on recently disturbed soil, especially building sites.

SIZE
Cap 3–7cm across. Stem 5–8cm high, 0.9–1cm thick.

MAIN IDENTIFICATION FEATURES
White, scaly, domed cap – hence the alternative name, Lawyer's Wig.

The white cap, with its buff centre, is at first egg-shaped above its swollen stem. As it matures it becomes cylindrical and then bell-shaped. The distinctive curling scales on the outer surface provide the mushroom's alternative name, Lawyer's Wig. The gills begin white, slowly changing to black from the edge inwards and finally dissolving into an inky mush. Only the young mushrooms are edible, and they need using quickly as they rapidly decay to a black sludge. Best collected when the gills are still white, just as the rim begins to emerge from the base of the egg-shaped cap.

Cooking suggestions
Use only young, fresh caps. They are good on their own or mixed with Parasol Mushrooms (*Macrolepiota procera*) (see page 82), in a simple and superb soup – just the mushrooms, some onions, and enough potato for thickening. Their Field-Mushroomy flavour is fine in sauces too.

Best storage method
Dried, but only with an electric drier (see *Storage Methods*, page 144).

> **WARNING**
>
> **Beware of its lookalike: the Common Ink Cap (*Coprinus atrmentarius*) (see page 128).**

Craterellus cornucopioides
Horn of Plenty

EDIBILITY Excellent

SEASON CHECK
July to November
(Summer to late autumn)

HABITAT
All types of woodland in glades. Hard to find, but can occur in large quantities.

SIZE
Cap 2–8cm across. No stem.

MAIN IDENTIFICATION FEATURES
Brown to grey-black and trumpet-shaped. Difficult to see among the leaf litter.

Like Chanterelles (*Cantharellus cibarius*) (see page 50), these mushrooms, which are mid-brown to black in colour, tend to appear in large groups in the same places every year, especially after autumn rain. Like Winter Chanterelles (*Cantharellus infundibuliformis*) (see page 52), they are well camouflaged by fallen leaves. The hollow cap has an ash grey outer surface and is trumpet-shaped. Its wide rim becomes irregular, crisped and wrinkled as it matures. The flesh is thin and leathery. This mushroom is no beauty, but has a fine flavour when cooked.

Cooking suggestions
This is one of the few mushrooms that needs good cleaning. Unlike many, it should be washed to remove grit from the trumpet. The rich, sweet, earthy flavour does wonders for soups, stews and casseroles. The dark colour doesn't fade with cooking, so these mushrooms look (and taste) particularly striking with white fish.

Best storage method
Dried after cleaning. Cut the larger caps in half first.

Fistulina hepatica
Beefsteak Mushroom

EDIBILITY Good

SEASON CHECK
July to November
(Summer to late autumn)

HABITAT
Almost exclusively on oak, but sometimes on Sweet Chestnut trees.

SIZE
Cap 10–25cm+ across, 2–6cm thick.

MAIN IDENTIFICATION FEATURES
Tongue-shaped red bracket on the main trunk or branches and, sometimes, stumps or fallen trees.

These mushrooms grow out of tree trunks, sometimes very high up. They are part of the group known as Bracket fungi. They form a kind of cushion, which grows into a tongue-shaped bracket. They are orange-red, growing purplish brown with age. Their succulent flesh is mottled with red and 'bleeds' a red sap when cut, making the mushroom look remarkably like raw meat (it is sometimes called Ox Tongue). Its upper surface can become very moist and spongy in wet weather. The individual tubes part very easily, unlike those of other bracket fungi.

Cooking suggestions
Use only when fresh. Separate the layers and wipe well, then slice into strips and soak them in milk for no less than a couple of hours, to remove any bitterness from the flavour. Then you can treat them like steak: grilled with onion, basil and garlic, perhaps, or barbecued. Added to soups and stews, they impart a richer flavour and colour.

Best storage method
Can be dried, but it is usually best to freeze them in pre-cooked dishes.

Flammulina velutipes
Velvet Shank

EDIBILITY Good

SEASON CHECK
November to March
(Late autumn to early spring)

HABITAT
Old tree-stumps and fallen branches, sometimes with gorse nearby.

SIZE
Cap 2–10cm across. Stem 1–3cm high, 0.4–0.8cm thick.

MAIN IDENTIFICATION FEATURES
Orange cap with dark brown velvety stem. Grows in dense clusters.

This mushroom, with its dark velvety stem, grows in dense clusters in winter and can withstand frosts: indeed, it does not usually appear until there has been a frost. The light-orange cap, darkening towards the centre, is smooth and shiny, and its surface is sticky or slimy to the touch. The cap is convex when young, then flattens out. The gills are pale yellow. The stem is yellow at the top and dark velvet brown below. The thin flesh has a pleasant smell and taste.

Cooking suggestions
This mushroom is quite tough, and therefore needs to be cooked thoroughly. Cut off the stems and slice finely. Cook them until they are tender. These mushrooms come into their own in soups and stews.

Best storage method
Dried.

WARNING

Beware of the Sulphur Tuft (*Hypholoma fasciculare*) (see page 132), which has a similar colour and grows in clusters, sometimes at the same time.

Grifola frondosa
Hen of the Woods

EDIBILITY Excellent

SEASON CHECK
September to November
(Autumn)

HABITAT
Bases of mature oak and beech trees, and sometimes stumps.

SIZE
Fruit body 15–40cm+ across. Cap 4–10cm across, 0.5–1cm thick.

MAIN IDENTIFICATION FEATURES
Grows in multiple layers at the tree base. Often very large. Occasionally found higher up the trunk.

This mushroom is relatively rare. It grows in clusters, mostly at the base of trees, but can sometimes be found some distance up the trunk. The clusters have a striking sculptural appearance. The caps, joined and overlapping, are olive-greyish (turning brown with age) with darker markings on the surface near the rim. At the rim each cap becomes feathery and wavy. The stem is cream to pale grey. The white flesh has a tough texture and a musty odour. Despite the off-putting smell, when the caps are young and fresh they taste very pleasant.

Cooking suggestions
Clean this mushroom very carefully because of all the nooks and crannies! Rinse in cold water before cooking, and cook well to get rid of the toughness. This is an excellent addition to any mushroom dish.

Best storage method
Dried, but also good in pre-cooked frozen dishes. After drying, it can be powdered for adding to soups and stews.

Hydnum repandum
Hedgehog Fungus

EDIBILITY Excellent

SEASON CHECK
August to November
(Late summer to late autumn)

HABITAT
Mixed woodland in damp areas around ditches. Likes beech particularly.

SIZE
Cap 3–17cm across. Stem 3.5–7.5cm high, 1.5–4cm thick.

MAIN IDENTIFICATION FEATURES
Pale buff to off-white cap, with spines rather than gills or pores.

This species is a prize for the mushroom hunter. In the right year it can be found in sizeable quantities. The small caps, ranging from pale creamy yellow to pinkish buff (and sometimes white) are flattened with a slight depression and a rolled rim. Instead of gills there are tiny spines on the underside of the cap – hence its name. The stout, tough stem is smooth and straight and can stain orange-brown when cut. The smell is pleasant, while the cooked flavour is excellent.

Cooking suggestions

After cleaning, small ones can be cooked whole or sliced. Remove the spines from larger specimens: though edible, they don't look so good on the plate! The young ones have a peppery, watercressy flavour that is excellent raw in salads. They are very versatile and good added to any dishes involving either meat or fish. Cooked simply in butter, they make a fine filling for pancakes.

Best storage method

Dried.

Handkea excipuliformis
Var. Puff-ball

EDIBILITY Good

SEASON CHECK
June to October
(Early summer to autumn)

HABITAT
Often waste ground, heaths and pastures. Also mixed woodland.

SIZE
Fruit body 3–10cm across, 8–20cm high.

MAIN IDENTIFICATION FEATURES
Light brown pestle-shaped cap with a sterile base.

This is another of the Puff-ball family, worth eating when it is young and firm with white flesh. The cap is shaped rather like a pestle and is a pale buff, getting browner as it ages. There are often small warts on the outer surface, but these disappear with age. The smell is mushroomy and pleasant.

Cooking suggestions
Cook as you would all other Puff-balls (and only pick the young ones) (see pages 47 and 49 for further details).

Best storage method
Dried.

Hygrocybe coccinea
Scarlet Hood

EDIBILITY Good

SEASON CHECK
August to November
(Late summer to late autumn)

HABITAT
Often grassy fields, but also beaches where Marram Grass is growing.

SIZE
Cap 2–4cm across. Stem 2–5cm high, 0.3–1cm thick.

MAIN IDENTIFICATION FEATURES
Bright red cap and stem, and a greasy, waxy texture.

This is one of the numerous Wax Caps, many of which are edible but not worth hunting for. The bright scarlet cap is a striking feature of this mushroom. Bell-shaped at first, the cap opens out when more mature. The flesh is yellowish red and hollow. The stem is the same colour as the cap, but has a yellowish base. The aroma is slight and not distinctive. This is a good mushroom for your basket, not only for its flavour but also for the colour it adds to dishes.

Cooking suggestions
This is good in all mixed mushroom dishes because of its colour and flavour. I have also found that it is very good blanched with other sweet mushrooms in spiced vermouth and served as a sauce with ice cream.

Best storage method
Dried. Alternatively, blanched and mixed with other sweet and colourful mushrooms, then stored in spiced alcohol for later use.

Hygrocybe pratensis
Meadow Wax Cap

EDIBILITY Good

SEASON CHECK
September to November (Autumn)

HABITAT
Pastureland. Also the edges of marshland.

SIZE
Cap 3.4–8cm across. Stem 2–5cm high, 1–1.5cm thick.

MAIN IDENTIFICATION FEATURES
Tawny-buff waxy cap. Often grows in a ring.

This is another of the Wax Caps that is worth collecting. The cap is tawny-buff and waxy to the touch. It is bell-shaped at first but soon opens out with a bump in the centre. The older caps often crack with age. Its aroma is typically mushroomy and its taste is pleasant. It can often be found in quite large rings in pasture – hence its name.

Cooking suggestions
Good in all mixed wild mushroom dishes.

Best storage method
Dried.

Laccaria amethystina
Amethyst Deceiver

EDIBILITY Good

SEASON CHECK
September to November (Autumn)

HABITAT
Mixed woodland. Particularly favours beech and Silver Birch.

SIZE
Cap 1–6cm across. Stem 4–10cm high, 0.5–1cm thick.

MAIN IDENTIFICATION FEATURES
Small bright lilac cap on a thin stem. Grows in large groups.

These mushrooms take some identifying as they can vary widely in appearance, which is how they got their name. They grow in large crowded groups. The cap is convex at first, then flattens and becomes wavy at the rim. The colour is a deep lilac throughout. The cap pales with age, but the hollow stem retains its deep colour. These are a good find: although they can take some time to gather, the taste is worth the effort.

WARNING

The *Mycena pura* (see page 136) is very similar and grows at the same time. It is more pink than lilac in colour. Although it is not deadly it can have side-effects for some people, so should be avoided.

Cooking suggestions
Due to its colour this is a good addition to all mixed mushroom dishes.

Best storage method
Dried, or stored in spiced spirit.

Laccaria laccata
The Deceiver

EDIBILITY Good

SEASON CHECK
August to November
(Late summer to late autumn)

HABITAT
Mixed woodland with beech and Silver Birch.

SIZE
Cap 1.5–6cm across. Stem 5–10cm high, 0.6–1cm thick.

MAIN IDENTIFICATION FEATURES
Small reddish brown cap on a thin stem. Grows in large groups.

This mushroom is similar in many respects to the Amethyst Deceiver (*Laccaria amethystina*) (see page 67) and it belongs to the same family. The Deceiver's cap is convex at first, then flattens out and becomes wavy at the rim. The colour is tawny to brick red and fades with age. The hollow stem retains its colour after the cap has faded. This mushroom is also found in large quantities.

Cooking suggestions
With its good colour and flavour, it is good in all mixed mushroom dishes.

Best storage method
Dried.

Lactarius deterrimus
Var. Saffron Milk Cap

EDIBILITY Good

SEASON CHECK
August to November
(Late summer to late autumn)

HABITAT
Open ground, often under spruce or other conifers.

SIZE
Cap 3–10cm across. Stem 3–6cm high, 1.5–2cm thick.

MAIN IDENTIFICATION FEATURES
Saffron-coloured cap with a distinct greenish tinge. Hollow stem.

In many ways this is very similar to The Saffron Milk Cap (*Lactarius deliciosus*) (see page 70). The main difference is a much greener tinge to the cap. The 'milk' is still orange, but after exposure to the air for about five minutes it gains a purplish tint, which darkens to a dull wine red.

Cooking suggestions
As for the Saffron Milk Cap (*Lactarius deliciosus*) (see page 70).

Best storage method
Dried.

WARNING

Be careful of the poisonous lookalikes, the Woolly Milk Caps (*Lactarius torminosus or Lactarius pubescens*) (see page 134): they both have hairy rims and are less bright than the Var. Saffron Milk Cap.

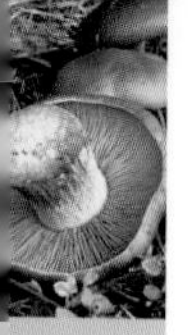

Lactarius deliciosus
Saffron Milk Cap

EDIBILITY Excellent

SEASON CHECK
August to November
(Late summer to late autumn)

HABITAT
Outside edges of coniferous woods.

SIZE
Cap 3–10cm across. Stem 3–6cm high, 1.5–2cm thick.

MAIN IDENTIFICATION FEATURES
Saffron yellow to orange cap, with a greenish tint. Concentric ring on cap. Gills bleed yellow-orange when cut.

This is an attractive and interesting mushroom, with its streaky carrot orange cap and clean in-rolled rim. The cap is convex with a depressed centre, fades with age to a pale silvery white or green, and bleeds saffron-coloured milk when cut. The saffron gills are closely packed and bruise green. The hollow stem is pale, blotched with orange, and green when broken or bruised. Pick the orange caps when young (but old enough to be distinguished from their poisonous lookalikes).

WARNING

Be careful of the poisonous lookalikes, the Woolly Milk Caps (*Lactarius torminosus or Lactarius pubescens*) (see page 134): they both have hairy rims and are less bright than the Saffron Milk Cap.

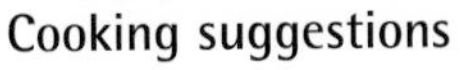

Cooking suggestions
Only use young fresh caps. As they often grow where there is sand present, you will need to wash them well. They have a good flavour, especially when fried with meat dishes – take advantage of their glorious colour and firm crunchy texture. Try them cooked in butter and served on toast, sprinkled with parsley and toasted pine nuts. These mushrooms are very popular on the continent and can often be found on market stalls.

Best storage method
Dried. Alternatively, stored in oil or vinegar after blanching.

Laetiporus sulphureus
Chicken of the Woods

EDIBILITY Excellent

SEASON CHECK
July to October
(Summer to autumn)

HABITAT
On mature oak but also occurs on beech and Sweet Chestnut. Often grows high up on the trunks (a telescopic pruning saw is a useful tool for picking these specimens).

SIZE
Bracket 10–40cm across, and can sometimes grow to 1m deep. (I have found one that weighed 50kg.)

MAIN IDENTIFICATION FEATURES
When young, the bright yellow brackets look rather like molten lava running down the host tree trunk. Older specimens are darker.

This is sometimes known as the Sulphur Polypore and is a really striking bracket fungus. It fans out in roughly semi-circular tiers from living tree trunks, and grows in large numbers. The eye-catching lemon yellow colouring darkens with age, when the flesh begins to get spongy, soft and crumbly. The succulent caps have a velvety or suede-like appearance and rounded edges. Their upper surface is uneven and lumpy, smoothing out a little with age. The smell is strong and fungus-like. Always pick young specimens that have a bright colour.

Cooking suggestions
Use only the young ones. Clean them thoroughly and blanch in salted water before cooking. The mushroom's intriguing texture and flavour (which is very similar to chicken), makes them a great substitute for poultry, when fresh and moist, in risotto or curry.

Best storage method
Can be dried, but I have found that it is best to pre-cook dishes and freeze them.

WARNING

A small percentage of people are allergic to this fungus, so be careful in serving it.

Leccinum quercinum
Var. Orange Birch Boletus

EDIBILITY Excellent

SEASON CHECK
July to October
(Summer to autumn)

HABITAT
Often with oak trees, but Silver Birch scrub is also a good hunting ground.

SIZE
Cap 6–15cm across. Stem 15–20cm high, 2–3.5cm thick.

MAIN IDENTIFICATION FEATURES
Chestnut-orange to brown cap, which is often large. Well-marked scaly stem.

This is one of the tastiest of the Boletus family. Its attractive chestnut-orange to date-brown cap is slightly scaly at first, becoming smooth later. The rim overhangs the pores. The stem is pale brown to buff, with whitish scales. This is one of the Leccinum, which have no marked changes when they are cut or bruised. It has a pleasant aroma and taste.

Cooking suggestions
This mushroom is a favourite among chefs because it softens when cooked and gives a superb texture to soups and casseroles. Try it in a cream sauce with fish, or mixed in paella, or in a varied wild mushroom dip.

Best storage method
Dried after being sliced (as the caps are often very large).

Leccinum scabrum
Brown Birch Boletus

EDIBILITY Good

SEASON CHECK
August to November
(Late summer to late autumn)

HABITAT
Mixed woodland, often (but not always) under birch.

SIZE
Cap 5–15cm across. Stem 7–20cm high, 2–3cm thick.

MAIN IDENTIFICATION FEATURES
Brown cap with buff-coloured pores, clothed with blackish scurfy scales.

The cap, a mid-brown snuff colour, is smooth and dry but gets rather tacky when wet. The substantial stem is white to grey with brownish-black scales, getting darker towards the base. The soft, watery flesh is white, sometimes flushing pink. The aroma is pleasant. Eat these when they are young and at their firmest.

Cooking suggestions
Even the younger, firmer ones have a soft texture, so they are at their best when added to mixed mushroom dishes, or in a creamy mushroom soup.

Best storage method
Dried. This helps to intensify its flavour.

Leccinum versipelle
Orange Birch Boletus

EDIBILITY Excellent

SEASON CHECK
August to November
(Late summer to late autumn)

HABITAT
Mixed woodland, especially where there are Silver Birches.

SIZE
Cap 8–20cm+ across. Stem up to 20cm high, 1.5–4cm thick.

MAIN IDENTIFICATION FEATURES
Bright orange cap. Striated stem, which bruises bluish green and is pinkish blue on cutting.

Another tasty Boletus mushroom. The striking tawny-orange cap, with its overhanging rim, is downy and fluffy when young but becomes smooth or scaly with maturity (depending on the weather conditions). The substantial white-grey stem has woolly brownish black scales. Young stems bruise in patches of bright electric blue. The flesh is pale, blackening with age, and bluish green at the stem base.

Cooking suggestions
Cook as recommended for the Var. Orange Birch Boletus (*Leccinum quercinum*) (see page 73).

Best storage method
Dried after slicing. It will go black with drying but this does not affect the flavour.

Lepista nuda
Wood Blewit

EDIBILITY Excellent

SEASON CHECK
October to December
(Autumn to early winter)

HABITAT
Often open ground or dappled shade with both coniferous and broad-leaved trees. Can also occur on golf courses and allotments.

SIZE
Cap 6–12cm across. Stem 5–9cm high, 1.5–2.5cm thick.

MAIN IDENTIFICATION FEATURES
Bright purple-lilac cap and stem.

Cortinarius camphorates

These are fine-flavoured mushrooms, but some people are allergic to them, so eat with caution, in small quantities and well-cooked – never raw. They are best picked when their colour is high. The bluish-lilac hue of the young mushrooms turns to a polished-looking tan brown, drying paler. The smooth cap is convex, flattening out and later becoming depressed and wavy. The tough, fibrous stem is slightly bulbous at the base and often has purple markings. The thick bluish flesh is strongly perfumed. The crowded lilac-violet gills fade to near-buff with age, but they do not turn brown.

Cooking suggestions

This mushroom must be cooked before eating. Its colour and fragrance make it effective in both sweet and savoury dishes. It has a strong flavour, so try pairing it with strongly-flavoured vegetables, such as onions or leeks, and then add a béchamel sauce.

WARNING

Be careful not to confuse this with the *Cortinarius camphorates* (above), which is similar in certain respects, although rare. Do also bear in mind that some people can be allergic to Blewits.

Best storage method

Good dried. Can also be stored after blanching in wine vinegar or extra virgin olive oil.

Lepista panaeolus (Lepista luscina)

EDIBILITY Good

SEASON CHECK
August to October
(Late summer to autumn)

HABITAT
Close, grazed, open grassland. Grows in rings.

SIZE
Cap 3–10cm across. Stem 3–5cm high, 0.4–0.6cm thick

MAIN IDENTIFICATION FEATURES
Dirty grey-brown cap. Grows in rings on open grassland.

This is another of the grassland mushrooms that is good for the basket. The caps are a dirty grey-brown in colour, convex at first, opening out as they grow. The flesh is greyish and the gills are white. Its mealy smell provides another identification feature. Care again must be taken not to confuse with the *Clitocybe dealbata and C. rivulosa* (see pages 126 and 127). With its nutty flavour and good texture it is a useful find.

> **WARNING**
>
> **Beware of this mushroom's lookalikes *Clitocybe dealbata and C. rivulosa* (see pages 126 and 127).**

Cooking suggestions
Due to the tasty flavour and texture, this makes a great addition to all mixed mushroom dishes.

Best storage method
Dried.

Lepista saeva
Field Blewit

EDIBILITY Excellent

SEASON CHECK
October to December
(Autumn to early winter)

HABITAT
Open fields, pasture and marshes, often after the first frosts.

SIZE
Cap 6–10cm+ across. Stem 3–6cm high, 1.5–2.5cm thick.

MAIN IDENTIFICATION FEATURES
Buff cap and purple striated stem.

This is sometimes called a Blue-leg or Blue-stalk, from its brightly-coloured bluish lilac stem (which may fade with age): this is the most obvious identification feature. The stem is sometimes rather bulbous at the base, with darker lilac markings. The cap is a shiny buff, which turns brown later. It is convex at first, then flattens and even depresses slightly, with a touch of waviness at the rim. The gills are whitish and crowded, and the flesh, white or flesh-coloured, is thick and chunky. Like the Wood Blewit (*Lepista nuda*) (see page 76-77), it can disagree with some people, so eat in small quantities and never raw.

Cooking suggestions
This must be cooked before eating. Chop and add to stews to show off its great flavour, which also enhances game, toasted nuts and strong cheese.

Best storage method
Dried but, like the Wood Blewit, it can also be stored after blanching, in white wine vinegar or extra virgin olive oil. This way you also then get the benefit of a flavoured oil, which can be used later.

WARNING

Be careful not to confuse with the *Cortinarius camphorates* (see page 77), which is similar in certain respects. Also bear in mind that some people can be allergic to Blewits.

Lycoperdon perlatum
Var. Puff-ball

EDIBILITY Good

SEASON CHECK
June to October
(Early summer to autumn)

HABITAT
Mixed woodland, but also in grassland. Often grows in very large groups.

SIZE
Cap 2.5–6cm across, 2–9cm high.

MAIN IDENTIFICATION FEATURES
Resembles a miniature cudgel with a warty top.

This mushroom is often very prolific and can only be eaten when fresh and young. The cap is white at first, becoming yellowish brown with age. It has an outer layer of tiny warts, which are especially dense on the top, and it has a rather distinct stem. I think this mushroom looks rather like a tiny club or cudgel with bobbles on the top! It has a pleasant mushroomy aroma.

Cooking suggestions
Only use young fresh caps. Like all Puff-balls it is good as a breakfast mushroom – particularly alongside bacon and eggs (see pages 47 and 49 for more details).

Best storage method
Dried.

Lyophyllum decastes

EDIBILITY Good

SEASON CHECK
August to October
(Late summer to autumn)

HABITAT
Open woodland often with Silver Birch.
Grows in clusters.

SIZE
Cap 4–10cm across. Stem 3–6cm high, 1.2cm thick.

MAIN IDENTIFICATION FEATURES
Shiny silvery grey-brown cap.
Occurs in clusters.

This mushroom grows in clusters in open woodland. The cap is convex at first, becoming wavy. It is light silvery grey to brown, has a shiny appearance and the flesh is white. The cluster can be quite large so this is a good find for your basket.

Cooking suggestions
This has a good flavour and texture and goes well in all mushroom dishes.

Best storage method
Dried.

Macrolepiota procera
Parasol

EDIBILITY Good

SEASON CHECK
August to October
(Late summer to autumn)

HABITAT
Often on roadside verges and with mature oak and beech (and on golf courses also).

SIZE
Cap 10–25cm across. Stem 15–30cm high, 0.8–1.5cm thick.

MAIN IDENTIFICATION FEATURES
Large cap, resembling a lady's sunshade in shape.

The long-stemmed Parasol can grow in large numbers and reappears in the same location each year, sometimes several times in one season. Its cap is spherical at first, then flattens out, but retains its 'opened parasol' look. It is pale buff in colour and marked by symmetrical patterns of coarse, shaggy scales. The flesh is thin, white and tastes sweet. Pick when dry – it can get very soggy when wet.

This large mushroom is easy to spot and identify. They often grow on the golf course near my home, and I can see them from my office window without needing binoculars.

Cooking suggestions

The stems are very tough, so discard them and only use fresh white caps. They have a rich, almost meaty, texture and flavour and are much favoured by vegetarians. Dust pieces of cap in seasoned flour and deep- or shallow-fry. Use a tempura-style batter to make a wonderful starter, served with a sweet chilli dipping sauce.

Best storage method

Dried. It can then be made into a good powder to add to soups and sauces – a wonderful store-cupboard standby.

WARNING

Be careful not to confuse with the *Lepiota cristata* (see page 135). Although very much smaller, it has similar cap markings.

Macrolepiota rhacodes
Shaggy Parasol

EDIBILITY Good

SEASON CHECK
August to October
(Late summer to autumn)

HABITAT
Often along hedgerows and small plantations, and often with Parasol mushrooms (*Macrolepiota procera*) (see page 82).

SIZE
Cap 5–15cm across. Stem 10–15cm high, 1–1.5cm thick.

MAIN IDENTIFICATION FEATURES
Large shaggy cap. Grows in groups beneath hedges.

This is similar in many respects to the Parasol, but it is usually smaller and its cap is more heavily marked with scales (hence its name). The stem is off-white with a pinkish brown tinge. The gills are white and later tinged with pink. The white flesh bruises reddish brown or pink and turns red when cut.

Cooking suggestions
Use in the same way as Parasol mushrooms (see page 82) – fried in tempura batter: they are every bit as good. Discard the stems and only use young caps.

Best storage method
Dried.

> **WARNING**
>
> Be careful not to confuse with the *Lepiota cristata* (see page 135). Although smaller, it has similar cap markings. Some people can also be allergic to this mushroom.

Marasimus oreades
Fairy Ring Champignon

EDIBILITY Good

SEASON CHECK
June to October
(Early summer to autumn)

HABITAT
The shorter grass of lawns, pasture and golf courses.

SIZE
Cap 2–5cm across. Stem 2–10cm high, 0.3–0.5cm thick.

MAIN IDENTIFICATION FEATURES
Cap rather like a fairy's bonnet. Light brown with a raised centre in the middle of the 'brim'.

An early summer mushroom, which grows in characteristic uneven circles (these can sometimes be as much as 100m across). Be careful not to confuse this mushroom with the highly toxic silky grey *Clitocybe rivulosa* (see page 127), which also grows in rings, close by and often at the same time.

The Fairy Ring is tan-coloured when young and moist, drying to a light buff, with a convex cap that retains a marked tan-tinged centre peak even after flattening. The gills are white to light brown and widely spaced. The flesh is thick and whitish, and has an aroma of fresh sawdust. You will often find large quantities of these, so it's a good idea to get down on your hands and knees with a pair of scissors!

Cooking suggestions
Brush clean – washing spoils the flavour. These are wonderful fresh, and very versatile, since they go well with meat, fish and mixed mushroom dishes. They have a lovely oaky scent, so try them on their own, cooked in sweet butter.

Best storage method
Dried. As an alternative, it can be stored after blanching, in spiced alcohol, wine vinegar or olive oil.

WARNING

Be sure not to confuse with the *Clitocybe rivulosa* (see page 127), which is deadly poisonous.

Meripilus giganteus
Giant Polypore

EDIBILITY Fair

SEASON CHECK
August to October
(Late summer to autumn)

HABITAT
Usually on beech or oak, often on old, cut, untreated stumps.

SIZE
Fruit body 50–80cm across, 1–2cm thick.

MAIN IDENTIFICATION FEATURES
Very large fan-shaped layers of caps with creamy white flesh.

As the name suggests, this is a very large fungus. It can only be eaten when young and fresh, as it becomes tough and inedible when older. The fruiting body consists of numerous flattened fan-shaped caps growing on top of each other. The cap is dark cream with fibrous scales and radially grooved with darker markings. The flesh is creamy white and soft with a pleasant smell. When cut or handled, the flesh turns black.

Cooking suggestions
This is treated in a very similar way to the Dryads Saddle (*Polyporus squamosus*) (see page 94). It needs careful cooking for some time, but also goes well in a mushroom mixture. Remember only to use the young specimens.

Best storage method
Dried and powdered for later use.

Morchella elata
Black Morel

EDIBILITY Excellent

SEASON CHECK
March to June
(Early spring to early summer)

HABITAT
Often coniferous woodland. Common in Scotland. Favours burnt ground and also chalky soils.

SIZE
Fruit body 5–12cm high.

MAIN IDENTIFICATION FEATURES
Reddish brown cap, which turns black with age. Cap also has a distinctive network of ridges and pits.

Mushrooms in the Morel family are much prized for their taste and versatility. The cap is hollow, narrow and conical. It is textured with ridges and pits in roughly parallel rows, and blackens with age. The stem is hollow and brownish in colour. The nooks and crannies of the cap make good hiding places for insects, so beware! Indeed, many chefs tend only to use dried specimens to be certain that they are bug-free. This is another springtime mushroom, so it is a good start to the season. It will also test your eyesight, being hard to find!

Cooking suggestions
This mushroom must be cooked and should never be eaten raw. Try baking the fruit bodies with a savoury stuffing. It makes a great accompaniment for meat and game dishes because of the strong flavour it imparts to a sauce. Always clean fresh Morels thoroughly to get rid of any hidden grit or lurking insects.

Best storage method
Dried.

WARNING

Beware of the False Morel (*Gyromitra esculenta*) (see page 130), which is similar in looks but does not have a hollow stem.

Morchella vulgaris
Common Morel

EDIBILITY Excellent

SEASON CHECK
March to June
(Early spring to early summer)

HABITAT
Open woodland on chalk. Coppiced woodland is good, as well as ground that has been burned. Also beaches where Marram Grass is growing.

SIZE
Fruit body 5–12cm high.

MAIN IDENTIFICATION FEATURES
Roughly egg-shaped cap with irregular pits and ridges, which pales with age.

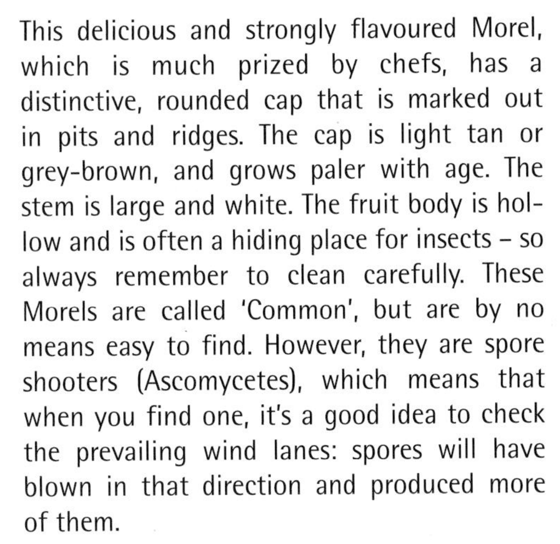

This delicious and strongly flavoured Morel, which is much prized by chefs, has a distinctive, rounded cap that is marked out in pits and ridges. The cap is light tan or grey-brown, and grows paler with age. The stem is large and white. The fruit body is hollow and is often a hiding place for insects – so always remember to clean carefully. These Morels are called 'Common', but are by no means easy to find. However, they are spore shooters (Ascomycetes), which means that when you find one, it's a good idea to check the prevailing wind lanes: spores will have blown in that direction and produced more of them.

Cooking suggestions
The same as for the Black Morel (*Morchella elata*) (see page 87). Because of its larger size, however, this is an ideal mushroom to cut in half and fill with stuffing. Always remember to clean this Morel well and remember that it should never be eaten raw.

Best storage method
Dried.

WARNING

Beware of the False Morel (*Gyromitra esculenta*) (see page 130), which is similar in looks but does not have a hollow stem.

Morchella esculenta
Var. Morel

EDIBILITY Excellent

SEASON CHECK
March to June
(Early spring to early summer)

HABITAT
Open woodland on chalk, but also on beaches where Marram Grass is growing.

SIZE
Fruit body 6–20cm+ high.

MAIN IDENTIFICATION FEATURES
Distinctive pits and ridges, which honeycomb the cap above a hollow stem.

Another member of the delicious Morel family, this mushroom is often quite large – and a great find in the spring, although you have to look hard for them. The cap is light brown, darkening with age, marked with ridges and pits with a roughly honeycombed texture. The stem is hollow, often swollen at the base and the flesh is white. As these mushrooms are spore-shooters (Ascomycetes) it is worthwhile, once you've found one, to look for more along the prevailing wind lanes – a tip that has stood me in good stead over the years.

Cooking suggestions
As for the Black and Common Morel (*Morchella elata* and *M. vulgaris*) (see pages 87–9).

Best storage method
Dried.

WARNING

Beware of confusion with the False Morel (*Gyromitra esculenta*) (see page 130).

Phallus impudicus
Stinkhorn

EDIBILITY Good

SEASON CHECK
July to November
(Summer to late autumn)

HABITAT
Likes rotting timber and occurs in all types of woodland, but especially coniferous.

SIZE
Egg 3–6cm across. Cap 10–25cm high (carried on the hollow white stem coming out of the egg).

MAIN IDENTIFICATION FEATURES
Unmistakeable, once you have seen the picture (and smelt the smell).

You can smell this fungus long before you see it. This clever mushroom has a bell-shaped cap at the top of a thick white stem, and the cap is covered with a brownish green slime which contains the spores. It is this that smells so revolting – and for a good reason: flies are attracted, settle on the cap, and fly away again carrying the spores, thus spreading the Stinkhorn far and wide. The only part that is edible for humans is the pinkish white egg-shaped sac (shown cut in half in the illustration), the fruit body out of which the stem rears. The egg tastes good, and is said to be an aphrodisiac – perhaps because of the mushroom's interesting appearance!

Cooking suggestions
Serve the egg raw. Only use the centre section, after removing the jelly. It tastes like a mixture between walnut and horseradish and is good in a dish of mixed mushrooms. (Don't show the picture to your friends until after they have eaten. We have found it to be a stimulating talking point!)

Best storage method
Only eat when freshly picked. If you just collect the 'eggs' and leave them, the Stinkhorn will emerge in a couple of days and you will need to move house!

The Stinkhorn in its egg form

Pleurotus cornucopiae
Var. Oyster Mushroom

EDIBILITY Good

SEASON CHECK
September to November
(Autumn)

HABITAT
Decaying trees and stumps, often with oak or beech, but predominantly elm. Their favourite host tree was elm, so these mushrooms have become harder to find since the loss of many trees to Dutch Elm Disease.

SIZE
Cap 6–14cm across. Stem 2–3cm high, 1–2cm thick.

MAIN IDENTIFICATION FEATURES
Creamy buff cap often fluted at the edge. Grouped in clusters by joined stems.

This is a very good mushroom to look for, as it is often found in very large quantities. The species grows in clusters with joined stems, and the clusters are usually in large groups. The cap is whitish buff and often fluted around the edge when larger. These mushrooms have a good flavour and are very versatile in the kitchen. Pick when young and fresh for the best results. They have been successfully grown commercially and are often available in a wide variety of colours that do not appear in the wild. In my opinion, however, the flavour of the commercial varieties is inferior to that of those found in the wild.

Cooking suggestions
A really useful mushroom with a delicate flavour and a delicious companion for all meats and fish. Excellent when served with slightly stronger mushrooms as well, and in sauces.

Best storage method
Dried, after separating the caps and slicing the larger ones.

Pleurotus ostreatus
Oyster Mushroom

EDIBILITY Good

SEASON CHECK
September to November (Autumn)

HABITAT
Decaying and fallen trees or tree stumps. Tends to favour elm, oak and beech.

SIZE
Cap 6–14cm across. Stem 2–3cm high, 1–2cm thick.

MAIN IDENTIFICATION FEATURES
Shell-shaped silver grey cap. Grows in clusters on fallen trees or stumps. (Some years ago we found a fallen beech covered in both types of Oyster Mushroom – about 200kg of them.)

The characteristic heaped clusters of shell-shaped wild Oyster Mushrooms grow in the same place year after year until their food source is exhausted. They are silvery grey, the caps later turning brownish. The rims of the larger caps may become fluted with age. The crowded gills are a creamy white, turning yellow as they mature. The stem is short and white. The flesh is also white and has a pleasant aroma.

Cooking suggestions
The same as for the Var. Oyster Mushroom (*Pleurotus cornucopiae*) (see opposite).

Best storage method
Dried, after separating the caps and slicing the larger ones.

Polyporus squamosus
Dryad's Saddle

EDIBILITY Fair

SEASON CHECK
March to August
(Early spring to late summer)

HABITAT
Deciduous trees – especially beech, elm and Sycamore. Often found on untreated stumps.

SIZE
Fruit body 5–65cm across, 0.5–5cm thick. Stem 3–10cm long.

MAIN IDENTIFICATION FEATURES
Saddle-shaped creamy-coloured cap with dark brown scales, joining directly onto the tree.

This mushroom is a bracket fungus and can only be eaten when it is young and tender. As the name suggests, it is like a saddle and grows in layers. These layers are circular and fan-shaped, while the fruiting body is dark cream in colour with dark brown concentric scales. They can grow to a very large size. The smell is mealy and pleasant. A common fungus, it grows over a long season, but you should only collect young, fresh ones. This is one of the parasitic fungi that often grow on stumps, eventually breaking them down completely.

Cooking suggestions
Its delicate flavour means it is ideal added to a mixture of mushrooms. It requires a little more time, so it is worth cooking it first and then adding other less dense mushrooms.

Best storage method
Dried and then powdered for later use in soups and stews and pasta dishes.

Russula claroflava
Yellow Swamp Russula

EDIBILITY Good

SEASON CHECK
June to November
(Early summer to late autumn)

HABITAT
Often under birch, on damp ground, as the name suggests.

SIZE
Cap 4–10cm across. Stem 4–10cm high, 1–2cm thick.

MAIN IDENTIFICATION FEATURES
Bright yellow and often shiny cap. White stem, which greys with age (or when bruised).

The Yellow Swamp Russula has a convex cap that is lemon yellow. It is fleshy, slightly sticky, and often shiny (though usually less so when dry). The margin is finely marked with lines. The stem is white and soft. The flesh is white and the gills are pale ochre, becoming creamy yellow with age. This mushroom tends to become dark grey with age or when bruised.

The Russula family is very large. There are over 100 different varieties and distinguishing them can be difficult. In this book, I have selected those that I have found most useful in the kitchen over the years. Many Russulas are edible but not worthwhile.

Some members of the Russula family are toxic: avoid all those that have red caps as these include the Sickener (*Russula emetica*) and the Beechwood Sickener (*Russula nobilis*) (see page 140).

Cooking suggestions
Very good in mixed wild mushroom dishes because of its crunchy texture, medium-to-hot taste and bright colour.

Best storage method
Dried. It provides a good dash of colour to your store of mixed dried mushrooms.

Russula cyanoxantha
The Charcoal Burner

EDIBILITY Good

SEASON CHECK
July to November
(Summer to late autumn)

HABITAT
They prefer beech. Also occur in mixed woodland.

SIZE
Cap 5–15cm across. Stem 5–10cm high, 1.5–3cm thick.

MAIN IDENTIFICATION FEATURES
Mottled purple-lilac or greenish brown cap. Flexible gills that don't break away easily.

Without doubt, this is the best of the Russula family, and you will have to beat the animals to secure this treat, as it's one of their favourite fungi. There are two distinct forms of this mushroom (both are illustrated). The cap of the first varies in colour: lilac, purple, wine red, olive green, or a mixture of colours. It may well show signs that insects have been at work, and is greasy to the touch when damp. The stem and flesh are white (sometimes flushed with purple), and the gills are creamy white. The other variety (*Russula cyanoxantha forma peltereaui*) has a greenish brown cap, also often marked by the ravages of insects. Both mushrooms are a good find, and with experience you will soon be able to spot them in the woods.

Cooking suggestions
Clean carefully – the mushrooms often need wiping with a damp cloth before use. They have a splendid flavour, and the texture is excellent: the mushroom stays quite crunchy even when cooked. Particularly good with meat and in mixed wild mushroom dishes.

Best storage method
Dried.

Russula cyanoxantha

Russula cyanoxantha forma peltereaui

Russula ochroleuca
Common Yellow Russula

EDIBILITY Good

SEASON
July to November
(Summer to late autumn)

HABITAT
Mixed broad-leaved woodland, as well as under conifers.

SIZE
Cap 4–10cm across. Stem 4–7cm high, 1.5–2.5cm thick.

MAIN IDENTIFICATION FEATURES
Yellow cap and white stem.

As its name suggests, this is a very common mushroom, pushing up through leaf litter throughout the season. The cap is muted yellow, convex at first, then flattening out and often soiled by the litter. You may want to collect only the clean ones, though the others are fine if cleaned thoroughly. The stem is white, greying with age and where bruised, and the gills are creamy white. The aroma is pleasant. Only collect the best specimens, as there will be plenty to choose from. They are a good find for your mixed basket, as they add colour and texture to your dishes.

Cooking suggestions
Just as for the Yellow Swamp Russula (*Russula claroflava*) (see page 95).

Best storage method
Dried.

Russula virescens

EDIBILITY Good

SEASON CHECK
July to November
(Summer to late autumn)

HABITAT
Under broad-leaved trees, especially beech.

SIZE
Cap 5–12cm across. Stem 4–9cm high, 2–4cm thick.

MAIN IDENTIFICATION FEATURES
Mould green to ochre-buff cap and cream-coloured stem.

Another good Russula for your basket. The cap is coloured a mottled dull green to yellow ochre. It is rounded when very young, then convex and later flattened, when it may often become wavy. A scaly scurf on the surface may develop. The stem is white and the gills a creamy white. This mushroom has a mild taste.

Cooking suggestions
Like all Russulas, this mushroom is good for both colour and taste, especially in mixed mushroom dishes.

Best storage method
Dried after careful cleaning and slicing.

Suillus bovinus

EDIBILITY Good

SEASON CHECK
July to October
(Summer to autumn)

HABITAT
Usually coniferous woodland, especially under Scots Pine.

SIZE
Cap 3–10cm across. Stem 4–6cm high, 5–8cm thick.

MAIN IDENTIFICATION FEATURES
Generally ochraceous or pinkish clay brown cap. Grows in clusters under pine.

This is another type of Boletus with pores and not gills. I have included several Suillus, as they are quite common and are good in mushroom cookery (especially after drying). They are much collected on the continent. The cap is yellow to cinnamon in colour and often quite shiny, with a distinct white margin. It is sticky to the touch. The stem is straw-coloured, darkening with age. The flesh is whitish to yellowish. The tubes are well spaced and the texture is quite open. The mushroom's aroma is pleasantly sweet.

Cooking suggestions
These are good in mixed mushroom dishes. They give off a lot of juice when cooking. This can be strained and used for the base of a sauce. I have found that all Suillus are best used from dried, as this intensifies their flavour.

Best storage method
Dried.

Suillus granulatus

EDIBILITY Good

SEASON CHECK
September to November (Autumn)

HABITAT
Usually coniferous woodland.

SIZE
Cap 3–9cm across. Stem 3.5–8cm high, 0.7–1cm thick.

MAIN IDENTIFICATION FEATURES
Shiny rust brown cap. Grows in clusters in coniferous woodland.

This is another variety of Suillus that is worth collecting. The cap is rust brown to yellowish and shiny when dry. The stem and flesh are lemon yellow. The tubes are yellow and quite dense. The aroma is slight and pleasant.

Cooking suggestions
These are good in mixed mushroom dishes – best dried first to intensify the flavour.

Best storage method
Dried.

Suillus grevillei
Larch Boletus

EDIBILITY Good

SEASON CHECK
September to November
(Autumn)

HABITAT
Almost exclusively under larch trees or on the edge of paths near larches.

SIZE
Cap 3–10cm across. Stem 5–7cm high, 1.5–2cm thick.

MAIN IDENTIFICATION FEATURES
Bright and shiny chrome yellow cap. Grows in clusters under or near larch trees.

This is a single-minded mushroom: it grows only in connection with larch trees. The cap is a striking yellow, which starts darker (with a rust-coloured flush) becoming brighter with age, and is shiny when dry. The yellow stem, encircled by a whitish ring, is often netted or otherwise marked. The flesh is light yellow in the cap but darker chrome in the stem.

Cooking suggestions
The same as for *Suillus bovinus* (see page 99). The striking colour of this mushroom makes it a useful addition to mixed mushroom dishes. They are best dried first.

Best storage method
Dried.

Suillus luteus
Slippery Jack

EDIBILITY Good

SEASON CHECK
July to October
(Summer to autumn)

HABITAT
Often around pine trees, but also along pathways on the forest edges.

SIZE
Cap 5–10cm+ across. Stem 5–10cm high, 2–3cm thick.

MAIN IDENTIFICATION FEATURES
Slippery brown cap with lemon yellow pores and upper stem.

This is also known as the Pine Boletus or Sticky Bun – the latter for obvious reasons. The cap is purplish to mid-brown, aging to a rusty orange-brown. It has an attractive sheen and is very sticky when wet (so pick in dry conditions, choosing only fresh mature specimens). The stem is pale yellow, often discoloured with darkening spots, and has a large ring, which is creamy white at first but darkens to purplish sepia. The flesh is initially yellow and turns brown with age. The aroma is pleasant. This is one of the largest Suillus and can sometimes be found in large quantities. It is sold in dried mixtures as the Yellow Boletus.

Cooking suggestions

Cook this on its own or with other Suillus, and strain the juice. Make a sauce with the juice, thicken it well and serve on toast with the mushroom and rashers of grilled or fried bacon. This mushroom softens well when cooked, so is excellent for soups. It is best to dry the mushrooms first.

Best storage method

Dried.

Suillus variegatus

EDIBILITY Good

SEASON CHECK
July to November
(Summer to late autumn)

HABITAT
Usually with conifers, sometimes on or beside forest paths.

SIZE
Cap 6–13cm across. Stem 5–9cm high, 1.5–2cm thick.

MAIN IDENTIFICATION FEATURES
Chunky rust brown cap with dense brown pores.

This Suillus is quite a chunky mushroom and is denser than many others of this type. The cap is rust brown, sometimes with an olive green tint, and is tacky in wet weather. The stem is yellow, with a rust-coloured flush at the base. The flesh is pale yellow in the cap. The pores are a darkish buff, becoming cinnamon-coloured, and quite dense.

Cooking suggestions
This is good in mixed mushroom dishes. It gives off less juice than the other Suillus, but can be cooked in the same way. It is best to dry the mushroom first.

Best storage method
Dried.

Sparassis crispa
Cauliflower Fungus

EDIBILITY Excellent

SEASON CHECK
August to October
(Late summer to autumn)

HABITAT
At the very base of pine trees, especially Scots Pine.

SIZE
Fruit body 20–50cm+ across. It can take several days to mature, so if you find a young one leave it for a few days (after covering with ferns to deter other seekers!) before picking.

MAIN IDENTIFICATION FEATURES
Resembles a very large cauliflower.

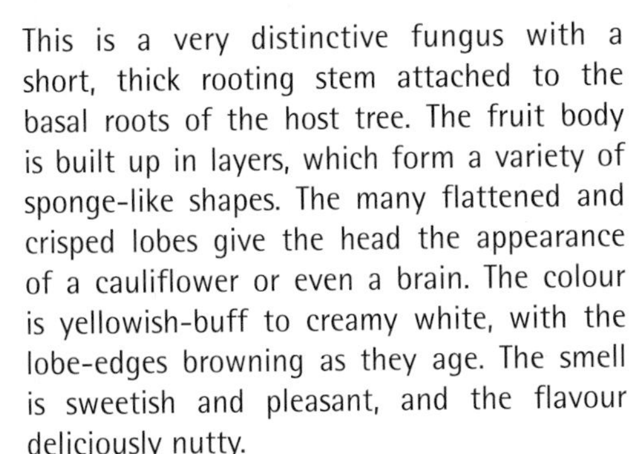

This is a very distinctive fungus with a short, thick rooting stem attached to the basal roots of the host tree. The fruit body is built up in layers, which form a variety of sponge-like shapes. The many flattened and crisped lobes give the head the appearance of a cauliflower or even a brain. The colour is yellowish-buff to creamy white, with the lobe-edges browning as they age. The smell is sweetish and pleasant, and the flavour deliciously nutty.

On our wedding anniversary – November 15th – Valerie and I were fortunate to find a specimen some 25kg in weight, which filled the boot of our car... Sadly, the tree it was attached to has since been felled.

Cooking suggestions
Use only the creamy white specimens. They can be washed, as the flesh is quite gelatinous. Break up the caps or slice them before cleaning, and always dry them thoroughly before use. Try them cut into thin slices (including the stem), dipped into a tempura batter and deep-fried: serve with a sweet chilli dipping sauce – delicious! They are also excellent sliced in stews and casseroles, or used in a fine cream and onion sauce.

Best storage method
Fresh whole specimens can be stored in a bucket of water for several days. For permanent storage, these are best cleaned very well and dried. They are often inhabited by wildlife: it's a good idea to hang them up for a day to allow the insects and bugs to depart!

TRUFFLE SPECIES

For centuries people have been intrigued by truffles: those seemingly magical, highly-flavoured gems from under the ground, which are not only difficult to find but are also very particular as to where they grow. The French and Italians have known about the wonderful scent and flavour of truffles for a long time. It is only relatively recently that their appeal has become worldwide. This is thanks to our ability to travel much more – and to the recent popularity of many TV food programmes extolling their culinary virtues. We now eat out more than ever, and increasing numbers of us have fallen under the spell of the truffle. Many people around the world await the truffle season eagerly in order to sample the superb flavour of these delicacies.

Truffle-collecting is a highly competitive activity, which requires the help of a dog or a pig to pick up their unique scent, as most of the time truffles grow under the soil. Dogs or pigs are traditionally used, because of their acute sense of smell. You'll not be surprised to know that nowadays most people choose to hunt with dogs: pigs are difficult to transport and at least two people are needed to handle them! The competition is so fierce these days and the price of the Alba Truffle (*Tuber magnatum*) (see page 109) so high, that there have been cases of dogs poisoned by rival hunters: a deplorable state of affairs. But it can't be denied: passions run very high during the season. There are even tales of shotguns being used to deter rivals on popular truffle grounds.

To the mushroom hunter, the truffle is the trophy that beats all others, simply because they are so hard to find. And once you have made a find you will do everything in your power to protect the site for future years. When I asked a mycological friend of mine how he found his first truffle, he told me it was by watching squirrels, which, along with deer, eat most of the UK's summer truffles. During the making of my video, *The Collector's Guide to Wild Mushrooms*, we actually filmed a

The secretive truffle hunter

The prize

squirrel digging up a black summer truffle. We followed his example and, I'm pleased to say, found some more.

Unless you have eaten truffles you will have no idea how delicious they are. If you've not tried them before, why not make this the year you have your first taste: a wonderful treat which makes any meal very special in the duller days of autumn and winter.

Maybe your first experience of truffles will be at a good restaurant, but you can make your own mouth-watering truffle dishes at home. Truffle oil is now readily available to buy, as well as truffles themselves. You can get them from specialist suppliers like *L'Aquila*, the largest supplier of truffles in the UK. You can also find them in the better-quality delicatessens and other specialist food stores. When travelling in France or Italy, look for truffles in the markets – but check the season first. These days you can even go on organised truffle-hunting holidays. They are great fun, but be warned: once you are hooked on the excitement and the taste of fresh truffles, you will want to go again and again!

In fact truffles are quite common in the UK, but we only have one variety: the Summer Truffle (*Tuber aestivum*) (see page 111). As I said, most of them are eaten by squirrels and deer.

The truffle is not an Ascomycete or spore-shooter, and so relies on being eaten by animals, which then pass on its spores through their droppings. A friend of mine in Bath dug up one of the trees in his garden last year, only to find 20 truffles attached to its roots. Needless to say the tree was replanted! He also found another 40 truffles under nearby trees, another stroke of good fortune for him. The animals had done a good job.

Truffles attach themselves directly to tree roots and are very choosy about which trees they like. So, as well as finding the right climatic conditions and soil type, you need to get to know the likely trees.

As you can imagine, there have been many attempts over the years to cultivate truffles. Some success has been achieved by planting acorns from infected oaks. The drawback: a 25-year wait while the roots spread and the truffles establish themselves. However, the process has been speeded up by inoculating hazels, which will produce truffles in a few years. These treated hazel trees can be bought from specialist suppliers. One Spanish farmer

Black Winter and White Truffles in the market

has cultivated a large forest with oaks carrying truffles. His investment over 25 years ago now produces a large quantity of them for the market. But dogs are still used to find them!

In this book, I have included details of the three most popular species. In order of importance, they are: (1) the Alba or White Truffle (*Tuber magnatum*) (see opposite); (2) the Black Winter or Perigord Truffle (*Tuber melanosporum*) (see page 110); (3) the Summer Truffle (*Tuber aestivum*) (see page 111).

The highly prized Alba Truffle commands the highest price of all: often as much as over £2,000 per kilogram, which per gram is more expensive than any other food at the time of writing. Its very intense flavour is unique, and it is only found in very specific areas.

The Black Perigord Truffle also has an excellent flavour and is highly prized. As its name suggests, it grows in the Perigord region of France and is often to be found in the local markets there. Although the price is lower, it still frequently exceeds £300 per kilogram.

The Summer Truffle has a good flavour, though it is not as intense. Summer Truffles usually cost about £100 or more per kilogram.

The truffle is supremely versatile in the kitchen. You can find more about cooking them at the end of each identification profile. Great care should be taken in cleaning and storing your finds, so as not to lose their wonderful aroma. I also find truffle oil a very important part of my cooking and always have it at hand in my store cupboard.

NOTE

There are a number of books about truffle hunting on the market. If you feel you might enjoy the challenge of the hunt, I suggest you research further before setting off on your own. Remember: this is a very secretive brand of hunting and hunters don't willingly pass on what they know.

Nature, however, can help with clues to the presence of truffles. The three main pointers are: (1) animal disturbance under a likely tree; (2) burnt ground around a tree; (3) a column of flies over a spot under a tree (this are only visible in still weather, though).

I wish you very happy truffle hunting – and do let me know if you have been successful... There are many truffles to find, if you have the patience to hunt them down. Good luck!

Tuber magnatum
Alba or White Truffle

EDIBILITY Excellent

SEASON CHECK
Late September to November, sometimes as late as January (Autumn to winter)

HABITAT
Mixed woodland, particularly favouring either hazel or poplar.

SIZE
Can vary widely. Fruit body up to 12cm across, 450g in weight, but most are around 5cm across and 30–50g in weight.

MAIN IDENTIFICATION FEATURES
Potato-like shape with smooth skin. Cream to light brown flesh, marbled with white veins. Very intense aroma. (This helps the animals find them – a trained dog can detect them at a distance of up to 50m.)

This is the most highly prized truffle of all. It grows in very limited areas and commands the highest price. The main growing region is Piedmont, of which Alba is the centre. This truffle has often been likened to a potato because of its colour and irregular shape. It has a smooth skin, and its flesh is hard and slightly brittle. Its aroma is very intense. As it grows beneath the ground, you will need either a pig or dog to find it.

Cooking suggestions
These are seldom cooked, as the process of cooking destroys some of their flavour. They should be very carefully cleaned and thinly sliced (using a truffle mandoline) over dishes, especially pasta.

Best storage method
They do not keep very long and must be kept in a closed container in a refrigerator – if the container is left open, all the food in the refrigerator will become truffle-scented! One good idea is to store them with dried rice as they will impart their flavour.

Another tip is to keep some eggs in a closed container with the truffles, so you can have truffle omelettes later. The shelf life varies but seven days is about the maximum. These truffles can be purchased in jars, professionally stored, but they do lack the full flavour of fresh specimens. There are many truffle products available, including truffle paste and truffle oil – expensive, but well worth it!

Tuber melanosporum
Black Winter or Perigord Truffle

EDIBILITY Excellent

SEASON CHECK
Mid-November to March
(Late autumn to early spring)

HABITAT
Usually with oaks, but is grown commercially on a variety of trees, often hazels.

SIZE
Fruit body up to 8cm across and 40–50g in weight. (Much larger specimens have been found, but these are rare.)

MAIN IDENTIFICATION FEATURES
Black, warty skin and pleasant aroma. It grows under the surface of the soil like all truffles.

This truffle comes from the Perigord and Provence regions of France. It is hailed by the French as the most important truffle – a claim naturally disputed by the Italians! It has an excellent flavour and intense aroma. The fruiting body is black and wart-covered. The flesh is very solid, marbled light brown with white veins, and turns black with cooking.

Cooking suggestions
This truffle is often cooked, but it can also be served raw – sliced after careful cleaning. The tough warty skin can be scrubbed carefully without loss of flavour. I have used these truffles very successfully to make my own truffle oil: it is wonderfully intense. A great addition to many dishes, including game, meat and pasta.

Best storage method
These truffles must be kept in a sealed container, and can be stored with dried rice so that it absorbs their flavour. They have a better shelf life than the Alba Truffle (*Tuber magnatum*) (see page 109) and can be refrigerated for up to 14 days. They are very versatile and are best used fresh.

Tuber aestivum
Summer Truffle

EDIBILITY Good

SEASON CHECK
June to November
(Early summer to late autumn)

HABITAT
Mixed woodland on chalky soil, particularly near beech and evergreen oak.

SIZE
Fruit body usually 3–4cm across and 20–30g in weight.

MAIN IDENTIFICATION FEATURES
Black warty shape, sometimes appearing just above the soil. Check for animal activity as a clue to their location.

This is the most widespread truffle. It grows extensively in the UK as well as on the continent. It has a warty appearance with a tough black skin. The flesh is creamy brown with white veins and the aroma is pleasant. This truffle sometimes breaks the surface of the soil, which can make it a little easier to find. However, unless you know the exact location, a trained dog will greatly assist the hunt. This truffle is less strong than the others and has a more subtle flavour.

Cooking suggestions
This can be used in a wide variety of dishes, including fish, due to its more delicate flavour. A very versatile truffle.

Best storage method
Store in a closed container, with rice or eggs which will pick up the flavour. Keep refrigerated. Shelf life is about 14 days.

POISONOUS SPECIES

Despite many warnings, people die of mushroom poisoning every year. These deaths stress how important it is to identify mushrooms correctly: it is literally vital that you take no chances when you are collecting.

Some people assume that, if animals or birds eat a mushroom, it must be edible. Unfortunately, they are wrong: for example, slugs can eat the lethal Death Cap *(Amanita phalloides)* (see page 118) with no ill-effects.

In this guide, I have provided descriptions of what I consider to be the poisonous mushrooms that all mushroom hunters should know and correctly identify. I have also given details of mushrooms that can be confused with edible species.

In each entry there is a detailed description of the mushroom. Next to its name is its classification of risk, from **Deadly Poisonous** down to **Dangerous** (for more information on the grades, see *How to Use This Book*, page 24). Then there follows information about the mushroom's growing season, habitat and size, and finally there are warnings about its poisoning effect.

The Amanita family of mushrooms pose the greatest threat: they include the Death Cap *(Amanita phalloides)* and the Destroying Angel *(Amanita virosa)* (see page 120), which are regarded as the most deadly. The toxin in any individual mushroom cap can vary, but an average-sized cap can contain enough toxins to kill at least 10 people. You must always remember that the poison from the Amanitas is easily transferred, so if you do accidentally handle one, wash your hands thoroughly before eating or handling other mushrooms. Wipes can be a useful addition to your equipment or a pair of plastic gloves which should be discarded after contact with the mushroom.

The most common poisoning is from the Yellow Stainer *(Agaricus xanthodermus)* (see page 115). It often grows with Horse Mushrooms *(Agaricus arvensis)* (see page 28) and Field Mushrooms *(Agaricus campestris)* (see page 30) and is quite often collected by mistake: so beware!

Fly Agaric (*Amanita muscaria*).

It's a good idea to have a camera with you on your mushroom hunts. Be sure to take photographs of any of the poisonous specimens for your files.

I do hope you won't find this section too worrying! The important thing is that it's always better to be safe than sorry. Remember the mushroom picker's golden rule:

IF IN DOUBT LEAVE IT OUT.

Agaricus semotus

POISONOUS

SEASON CHECK
August to October
(Late summer to autumn)

HABITAT
Open grassland, often alone or in small groups. Also the outside edges of woodland.

SIZE
Cap 2–5cm across. Stem 3–6cm high, 0.4–0.8cm thick.

MAIN IDENTIFICATION FEATURES
Yellowish brown cap with almost lilac-tinged scales in the centre of the cap.

This is quite a rare mushroom but can be confused with the other edible members of the Agaricus family, so care must be taken in correct identification of all that you collect.

The cap is white at first and then is covered in small scales in the centre of the cap. It has a distinct ring and is often tinged with yellow at the very base of the stem. The gills are pale at first turning pink and eventually grey. It smells mildly of aniseed.

POISONING EFFECT
Stomach cramps, sweating, flushing and sickness. Thankfully the symptoms are short-lived and not fatal.

WARNING:

AGARICUS SEMOTUS CAN BE CONFUSED WITH EDIBLE MEMBERS OF THE AGARICUS FAMILY (SEE PAGES 28–32).

Agaricus xanthoderma
Yellow Stainer

POISONOUS

SEASON CHECK
July to October
(Summer to autumn)

HABITAT
Widespread in woodland, hedgerows, grassland and marshes, and also gardens.

SIZE
Cap 5–15cm across. Stem 5–15cm high, 1–2cm thick.

MAIN IDENTIFICATION FEATURES
Instant yellow stain when bruised. Extreme base bright chrome yellow when cut. Unpleasant carbolic aroma. Grows in groups, often with Horse Mushrooms (*Agaricus arvensis*) (see page 28).

It is the collectors of Horse Mushrooms and Field Mushrooms (*Agaricus campestris*) (see page 30), who are most at risk from the Yellow Stainer. Around 50 per cent of the people who eat the species are adversely affected. It is also the cause of most mushroom poisoning by mis-identification.

Yellow Stainer caps are convex at first, then flatten out with a dip in the centre. The young caps are very white, but acquire small grey-brown scales as they mature. The flesh-coloured gills darken to grey-brown with age. Stems are white and bulbous, and stain chrome yellow at their bases.

The Yellow Stainer bruises a bright chrome yellow when touched or cut. It also has an unpleasant inky, carbolic smell. If you collect one in error and cook it, everything cooked with it may turn a sickly yellow – and give off an even stronger bitter smell.

WARNING:

DO NOT CONFUSE THIS WITH THE FIELD MUSHROOM (*AGARICUS ARVENSIS*), HORSE MUSHROOM (*A. CAMPESTRIS*) OR WOOD MUSHROOM (*A. SILVICOLA*) (SEE PAGES 28, 30 AND 31).

POISONING EFFECT
Stomach cramps, sweating, flushing and sickness. Thankfully the symptoms are short-lived and not fatal.

Amanita citrana var. citrina
Amanita citrina var. alba
False Death Cap

DANGEROUS

SEASON CHECK
August to November
(Late summer to late autumn)

HABITAT
Mixed woodland. *A. citrina var. alba* often found by beech or Silver Birch.

SIZE
Cap 4–10cm across. Stem 6–8cm high, 0.8–1.2cm thick.

MAIN IDENTIFICATION FEATURES
Ring on stem and large volval bag at the bulbous base. Lemon-yellow *A. citrina var. citrina* caps; white *A. citrina var. alba* caps. Both usually scattered with veil fragments.

Although this mushroom is not poisonous, I have included it in this section because it is easy to confuse with the highly deadly Death Cap (*Amanita phalloides*) (see page 118).

There are two forms of this mushroom: *Amanita citrina* and *Amanita citrina var. alba*. *Amanita citrina* caps are ivory to pale lemon, but *Amanita citrina var. alba* caps are pure white (and so are the flesh and stems). False Death Caps are usually covered with coarse whitish patches (remnants of the embryonic veil) which darken to ochre-brown and distinguish them from the true Death Cap – however, these fragments may get washed off by rain. The stems are ivory white and tapering, with a large sac at the base. The off-white gills darken with age. They have an unappealing aroma rather like raw potatoes. They are common during the main season.

WARNING:

THIS MUSHROOM CAN BE CONFUSED WITH THE DEATH CAP (*AMANITA PHALLOIDES*) (SEE PAGE 118).

Amanita muscaria
Fly Agaric

POISONOUS

SEASON CHECK
August to November
(Late summer to late autumn)

HABITAT
Widespread in woodland coppices, often near Silver Birch.

SIZE
Cap 8–20cm across. Stem 8–18cm high, 1–2cm thick.

MAIN IDENTIFICATION FEATURES
Bright red cap, flecked with white veil remnants (although these may be washed off in rain).

This is the storybook toadstool, with its pale-flecked scarlet cap (though those warty flecks may be washed off by rain). Hemispherical at first, the caps flatten out and the bright colour may begin to fade. The gills are white. The white stem has a bulbous base, revealed when the mushroom is lifted out of the ground.

Take care not to confuse this with the Caesar's Mushroom (*Amanita caesarea*) (see page 34). Although the Caesar's Mushroom does not yet grow in the UK, it is common and much sought after in Southern Europe. The Caesar's Mushroom has a large sac at the stem base, the cap is more orange in colour, there are no pale flecks, and the stem is yellow (shown clearly when cut in half).

POISONING EFFECT
Poisons the central nervous system, causing hallucinations and stupor. (It was commonly used to poison flies, which is how it got its name.)

Collectors' tip
Remember that the Fly Agaric can act as a flag mushroom for an edible species: it grows in similar ground and at the same time as the Cep (*Boletus edulis*) (see page 44).

WARNING:

THIS MUSHROOM CAN BE CONFUSED WITH CAESAR'S MUSHROOM (*AMANITA CAESAREA*) (SEE PAGE 34).

Amanita phalloides
Death Cap

DEADLY POISONOUS

SEASON CHECK
August to November
(Late summer to late autumn)

HABITAT
Widespread in mixed woodland, especially near oak and beech.

SIZE
Cap 4–12cm across. Stem 5–13cm high, 1–1.8cm thick.

MAIN IDENTIFICATION FEATURES
Sickly green cap, turning brown with age. Large sac at the base of the stem.

This is the most deadly fungus of all: just one can kill many people and there is no known antidote. The Death Cap can vary quite a lot in its appearance: if you are in any doubt at all, do not pick. It is best not to handle the Death Cap for fear of transmitting the toxins to your

WARNING:

TAWNY GRISETTE (*AMANITA FULVA*) (SEE PAGE 35) CAN BE CONFUSED WITH THIS DEADLY MUSHROOM.

edible mushrooms. If you handle one, however, always ensure that you wash your hands and your knife, if you have cut them. Wipes can be useful for this purpose – remember to dispose of them carefully after use.

The greenish caps are rounded at first, but flatten out later and turn a grubby-looking brown (though they may grow paler in wet weather, when they also become shiny). They often have a streaked, fibrous look. The gills are white and crowded at first, becoming quite open with age. The stem is white; the flesh is also white, with a yellow tinge near the cap. The aroma is sickly-sweet and increases after picking. There is a substantial sac at the base – a good identification aid – but you will nearly always need to clear away leaf mould and debris around the stem to find it: be sure to look.

POISONING EFFECT

Symptoms first appear some 6–24 hours after ingestion, by which time the toxins have already attacked the cells of the liver and kidneys. Symptoms include severe sickness and diarrhoea, accompanied by acute abdominal pains, lasting for a few days. There is then a period of apparent recovery, but death occurs several days later from total liver and kidney failure. There is, at present, no antidote, but if the source is identified within a few hours of the first symptoms, success may be achieved by blood transfusions.

Amanita virosa
Destroying Angel

DEADLY POISONOUS

SEASON CHECK
August to November
(Late summer to late autumn)

HABITAT
Mixed or deciduous woodland, often near oak and beech.

SIZE
Cap 5–12cm across. Stem 9–12cm high, 1–1.5cm thick.

MAIN IDENTIFICATION FEATURES
Pure white cap. Baggy, flowing volval sac at the base.

This is another of the deadly Amanitas. Be careful never to touch a Destroying Angel without wearing gloves: its poison is very easily transferred and is deadly. If handled by mistake, wash your hands and your knife (if you have used one), or use wipes and then carefully dispose of them.

This mushroom has white gills and a pure white cap without veil remnants. Its deceptively innocent appearance makes it easy to mistake for a young Field Mushroom (*Agaricus campestris*) (see page 30), so be absolutely sure you get the identification right. The cap is conical at first, but later becomes bell-shaped. There is often a distinct blunt point at the centre. The rim becomes irregular as it opens out. The crowded gills are also pure white. The white stem, which grows out of a large greyish sac at its base, has a shaggy, fibrous surface and is often slightly curved. The Destroying Angel has a sickly-sweet smell.

POISONING EFFECT

Exactly the same as the Death Cap (*Amanita phalloides*) (see page 118–119).

WARNING:

BEWARE OF THIS INNOCENT-LOOKING MUSHROOM. IT IS DEADLY AND CAN BE MISTAKEN FOR THE FIELD MUSHROOM (*AGARICUS ARVENSIS*) (SEE PAGE 30).

Amanita pantherina
Panther Cap

DEADLY POISONOUS

SEASON CHECK
August to November
(Late summer to late autumn)

HABITAT
Deciduous and coniferous woodland, and in beech woodland on limestone soil. Not common.

SIZE
Cap 6–10cm across. Stem 9–13cm high, 1–1.5cm thick.

MAIN IDENTIFICATION FEATURES
Bronze cap with white veil remnants. Often a double ring at the base of the chunky white stem.

This rare mushroom has a bronzed cap, scattered with small white patches – remnants of the embryonic veil (which also clings to the rim). The gills are crowded and white. The rather chunky white stem has a narrow hoop-like ring, fairly low down and rather ragged, and there are sometimes more rings just above the bulbous base. The stem hollows with age. The flesh is white.

POISONING EFFECT
Very similar to that of the Death Cap (*Amanita phalloides*) (see page 118–119). Poisoning occurs several hours after ingestion and is severe or fatal. The toxins attack the liver and kidneys, which are likely to fail altogether.

Panther caps are very poisonous and are deadly. *Amanita pantherina* is a close relative of the edible Blusher (*Amanita rubescens*) (see page 36) and slightly similar in looks (this goes especially for the young caps), so great care must be taken in identifying these species.

WARNING:

THIS DEADLY MUSHROOM CAN BE CONFUSED WITH THE BLUSHER (*AMANITA RUBESCENS*) (SEE PAGE 36).

Boletus erythropus

DANGEROUS

SEASON CHECK
July to November
(Summer to late autumn)

HABITAT
Coniferous, broad-leaved and mixed woodland. Often grows in small groups.

SIZE
Cap 8–20cm across. Stem 5–14cm high, 2–5cm thick.

MAIN IDENTIFICATION FEATURES
Snuff brown cap with yellowish rim, orange pores and red stem. Turns blue-black when bruised or cut.

Although collected and eaten by some people, this mushroom can cause nasty gastric upsets. It is also very easily confused with *Boletus satanoides* and Satan's Boletus (*Boletus satanas*) (see pages 124 and 125), both of which are poisonous and should never be collected.

The cap is snuff brown, becoming yellowish towards the rim. It has a slightly velvety surface, which grows smooth with age (and may be sticky when wet), and bruises blue-black. The stem is yellow and covered with dense red dots in a lattice pattern. The tubes are lemon yellow, becoming greenish, and the pores are orange-red, growing rust-coloured with age. Both tubes and pores turn dark blue when bruised or cut and the flesh is yellow, but quickly turns dark blue when cut. The colour changes are a good indentification feature of this mushroom.

POISONING EFFECT
This mushroom causes gastric upsets in many people.

WARNING:

IF YOU ATTEMPT TO EAT THIS MUSHROOM, IT MUST BE COOKED. MY ADVICE IS TO LEAVE IT ALONE, JUST IN CASE. IT IS ALSO VERY SIMILAR TO POISONOUS BOLETES, SUCH AS *BOLETUS SATANOIDES* (SEE PAGE 124) AND SATAN'S BOLETE (*BOLETUS SATANAS*) (SEE PAGE 125).

Boletus satanoides

POISONOUS

SEASON CHECK
June to September
(Early summer to early autumn)

HABITAT
Usually associated with oak trees.

SIZE
Cap 5–14cm across. Stem 8–16cm high, 2–4cm thick.

MAIN IDENTIFICATION FEATURES
Slightly shiny coffee-coloured cap. Thick red stem with a network of markings. Red to orange pores. Flesh turns bluish when bruised.

Poor-quality specimens are often found.

The name alone should put you off this poisonous mushroom! Although it is quite rare, it is one you should identify carefully. The cap is dirty white to begin with, turning to a light coffee colour with age and acquiring a reddish flush. There is a nice sheen on the cap's surface, which is often crazed. The bulbous stem is orange-red, becoming yellower towards the base, and has an orange-red network of markings. The pores are red, turning to a darker reddish orange with age. The flesh is white to pale lemon in colour, and turns slightly blue when cut or bruised. The aroma is spicy.

WARNING:

BEWARE OF ALL BOLETUS MUSHROOMS WITH ORANGE-RED PORES AND RED STEMS.

POISONING EFFECT
Sickness and stomach cramps. Thankfully the results are not fatal.

Boletus satanas
Satan's Boletus

POISONOUS, POTENTIALLY DEADLY POISONOUS

SEASON CHECK
July to September
(Summer to early autumn)

HABITAT
Broad-leaved trees, especially oak or beech, usually on chalk or limestone soil.

SIZE
Cap 8–25cm across. Stem 5–8cm high, 5–11cm thick, but tapering quickly.

MAIN IDENTIFICATION FEATURES
Rounded white-to-buff cap with very red pores. Very bulbous stem.

Another warning name! The Satan's Boletus is the most dangerous of the poisonous Boletus mushrooms, but happily it is rare. There are also several others (also rare) in this group, which should be avoided – all of them have orange to red pores and reddish stems.

The cap is white to buff with a red flush at the margins, with small cracks appearing on the surface, which turns brown where bruised or even just handled. The stem is short and usually bulbous, sometimes looking rather like a miniature turnip. The stem is yellowish, covered with a red network of tiny dots. The pores are red at the centre and become orange towards the margin. They bruise a greenish colour. The flesh is yellow (but paler or even white in the stem) and becomes blue when cut or bruised. The smell is unpleasant.

POISONING EFFECT
Sickness and stomach cramps, but the results are usually not fatal. Sometimes, however, Satan's Boletus can be fatal because the amount of toxins in each cap varies widely.

WARNING:

BEWARE OF ALL BOLETUSES WITH ORANGE-RED PORES AND RED STEMS. IF THEY ARE RARE VARIETIES, LIKE THIS ONE, PHOTOGRAPH THEM FOR YOUR RECORDS, BUT OTHERWISE LEAVE THEM WELL ALONE.

Clitocybe dealbata

DEADLY POISONOUS

SEASON CHECK
July to October
(Summer to autumn)

HABITAT
Lawns, pasture and along woodland paths.

SIZE
Cap 2–5cm across. Stem 2–3cm high, 0.5–1cm thick.

MAIN IDENTIFICATION FEATURES
Small off-white caps. Crowded whitish gills. Slight mealy smell. Often grows in groups and sometimes in rings.

Blending into the crowd, this small, deceptive and deadly mushroom grows at the same times and in the same places as edible mushrooms. Be sure you do not include it in your basket by mistake.

The caps are off-white to buff in colour, often with a dusting of white which makes them look slightly frosted. They are thin and flat with a thicker central depression, and the rims are somewhat in-rolled and fluted. Crowded cream-coloured gills run down the whitish stems. The smell is slightly mealy.

POISONING EFFECT
Attacks main organs, such as the liver and kidneys, with fatal consequences.

WARNING:

THIS DEADLY MUSHROOM CAN BE CONFUSED WITH THE MILLER (*CLITOPILUS PRUNULUS*) (SEE PAGE 56) AND *LEPISTA PANAEOLUS* (SEE PAGE 78). BE VERY CAREFUL NOT TO PICK IT BY MISTAKE.

Clitocybe rivulosa

DEADLY POISONOUS

SEASON CHECK
July to October
(Summer to autumn)

HABITAT
Grassland on sandy soil. Often beside paths and roads. Tendency to grow in rings.

SIZE
Cap 2–4cm across. Stem 2–3.5cm high, 0.5–1cm thick.

MAIN IDENTIFICATION FEATURES
Small grey-white caps. Crowded whitish gills. Grows in troops or rings, often near Fairy Ring Champignons (*Marasimus oreades*) (see page 85).

Take care when you are collecting the Fairy Ring Champignon – *Clitocybe rivulosa* grows at the same time and in the same places, often close by. Examine examples of both, very carefully, to ensure that you know which is which.

The smooth, silky caps begin cup-shaped, then flatten out with a small depression in the centre. They are greyish, powdered white, and may have concentric rings or marks where the flesh-coloured gills show through. They may also darken to light brown when wet. The gills are crowded and whitish buff, running down the stem. (The gills of Fairy Ring Champignon, by contrast, are thick and widely spaced.) The stems of *Clitocybe Rivulosa* are the same colour as the caps, with the flesh a grubby whitish grey or buff. The aroma is sweet but the mushroom is deadly.

POISONING EFFECT
Attacks main organs, such as the liver and kidneys, with fatal consequences.

WARNING:

THIS MUSHROOM CAN BE CONFUSED WITH THE HIGHLY-PRIZED FAIRY RING CHAMPIGNON (*MARASIMUS OREADES*) (SEE PAGE 85), THE MILLER (*CLITOPILUS PRUNULUS*) (SEE PAGE 56) AND *LEPISTA PANAEOLUS* (SEE PAGE 78). BE CAREFUL!

Coprinus atramentarius
Common Ink Cap

DANGEROUS

SEASON CHECK
July to November
(Summer to late autumn)

HABITAT
Decaying logs or tree stumps, or where wood is buried.

SIZE
Cap 3–7cm across. Stem 7–17cm high, 0.9–1cm thick.

MAIN IDENTIFICATION FEATURES
Bell-shaped cap. Thin white to light brown stem. Dissolves into an inky sludge. Grows in tufts, often close together.

Although this mushroom is edible, it has alarming effects if consumed in conjunction with alcohol, and has thus been used as a cure for alcoholism (as an antabuse).

This mushroom can be distinguished from the Shaggy Ink Cap (*Coprinus comatus*) (see page 57) by its lack of shaggy scales and the fact that its cap is dirty white instead of white. The cap is bell-shaped, often with fragments of the embryonic veil clinging to the surface, and its rim tends to pucker and split. The crowded gills are white at first, growing brown and finally disintegrating into an inky mass from the edge inwards. The stem is white and smooth with a brown scaly base. These mushrooms grow in tufts, which are often numerous and close together.

WARNING:

THIS MUSHROOM CAN BE MISTAKEN FOR THE SHAGGY INK CAP (*COPRINUS COMATUS*) (SEE PAGE 57). IF CONSUMED, DO NOT COMBINE IT WITH ALCOHOL.

POISONING EFFECT
Results are instant and unpleasant, and include sickness and sweating. After the mushroom is ingested, alcohol can continue to trigger its effects for up to four days.

Collectors' Tip
The decayed Common Ink Cap mushroom makes very good drawing ink. The species was used by mediaeval monks.

Coprinus picaceus
Magpie Fungus

POISONOUS

SEASON CHECK
July to November
(Summer to late autumn)

HABITAT
Usually on alkaline soil in conjunction with beech trees.

SIZE
Cap 3–7cm across. Stem 5–8cm high, 0.9–1cm thick.

MAIN IDENTIFICATION FEATURES
Unusual dark bell-shaped cap with whitish veil remnants still attached.

This slightly unusual mushroom with its slowly decaying veil has an almost magpie-like colouring, hence its name. Many people suffer internal upsets if they eat it, so it is best avoided altogether.

The conical white cap becomes bell-shaped and grey-black with age. White or pinkish veil remnants remain scattered on the cap's surface. The longish white stem has a woolly bulbous base. Like its Coprinus cousins, the Magpie eventually dissolves into an inky mass.

POISONING EFFECT
Disagreeable internal disorders, such as sickness and sweating.

Collectors' Tip
These can often be found where Horn of Plenty (*Craterellus cornucopioides*) (see page 58) are nearby, so may act as a flag mushroom for them.

WARNING:

THIS MUSHROOM UPSETS MOST PEOPLE, SO LEAVE IT ALONE. DO NOT CONFUSE WITH THE SHAGGY INK CAP (*COPRINUS COMATUS*) (SEE PAGE 57).

Gyromitra esculenta
False Morel

DEADLY POISONOUS

SEASON CHECK
March to May
(Spring)

HABITAT
Usually with conifers on sandy soil.

SIZE
Cap 3–9cm across. Stem 3–6cm high, 1–2cm thick.

MAIN IDENTIFICATION FEATURES
Crumpled, convoluted and irregular reddish-brown cap.

This is a mushroom not to be confused with those of the Morel family (see pages 87–90). It grows in the spring, as do they.

The cap of the False Morel is dark reddish brown and irregularly lobed. It looks a little like a misshapen brain. The short stem is flesh-coloured, faintly grooved and becomes hollow with age.

POISONING EFFECT
Deadly poisonous if eaten raw. Some people say that you can eat the mushroom after blanching in boiling water (being sure to discard the water afterwards). However, current research shows that the toxins that the mushroom contains build up in the system, attacking the liver and kidneys with potentially fatal consequences.

WARNING:

TAKE CARE NOT TO CONFUSE THIS POISONOUS MUSHROOM WITH EDIBLE MORELS (SEE PAGES 87–90). COMPARE THE TWO AND THE DIFFERENCE QUICKLY BECOMES CLEAR: EDIBLE MORELS DO NOT HAVE THE RANDOMLY CRUSHED APPEARANCE OF THE FALSE MOREL.

Hygrophoropsis aurantiaca
False Chanterelle

POISONOUS

SEASON CHECK
July to November
(Summer to late autumn)

HABITAT
Woodland and heathland, often with coniferous trees.

SIZE
Cap 2–8cm across. Stem 3–5cm high, 0.5–1cm thick.

MAIN IDENTIFICATION FEATURES
Orange colour. Gills running into the thin stem. Often found in troops close to true Chanterelles (*Cantharellus cibarius*) (see page 50).

Like many poisonous mushrooms, the False Chanterelle grows close by its edible lookalikes. Fortunately this one isn't deadly, but it causes gastro-intestinal problems and hallucinations, and therefore should be avoided.

The velvety caps are flat and in-rolled when young, growing funnel-shaped, then fluted like the true Chanterelles. However, they are orange-yellow in colour – not true Chanterelle yellow. The soft gills are dark orange, closely packed, sometimes forked, and run down the stem. The stems are the same colour as the caps, or might be slightly darker and are often curved. There is a distinct mushroomy aroma.

POISONING EFFECT
Severe digestive upsets and other unwelcome symptoms. Should be avoided.

WARNING:

DO NOT CONFUSE WITH THE CHANTERELLE (*CANTHARELLUS CIBARIUS*) OR WINTER CHANTERELLE (*C. INFUNDIBULIFORMIS*) (SEE PAGES 50 & 52).

Hypholoma fasciculare
Sulphur Tuft

POISONOUS

SEASON CHECK
All year round

HABITAT
Decaying timber and tree stumps, both deciduous and coniferous.

SIZE
Cap 2–7cm across. Stem 4–10cm high, 0.5–1cm thick.

MAIN IDENTIFICATION FEATURES
Caps that are, at first, yellowish tan, then sulphurous with an increasingly green tinge. Thin stems, often curved. Grows in clusters.

This poisonous mushroom is very common and can be found throughout the year, even in winter. It grows in clumps from dead or dying wood, and therefore can be confused with edible mushrooms, such as the Honey Fungus (*Armillaria mellea*) (see page 37) or Velvet Shank (*Flammulina velutipes*) (see page 60). Check your identification with care.

The caps are convex, often with yellow veil fragments like a tacking stitch around the edge, which can become black with spore deposit. They are bright yellow with a darker orange-tan centre. The gills are bright yellow at first, darkening to a sulphurous green and then brown. The flesh is bright yellow too, but brownish at the base of the stems, which are often curved and have purplish brown spores deposited at the top. There is a distinct mushroomy aroma.

POISONING EFFECT
Can cause stomach upsets and other related symptoms. Should be avoided.

WARNING:

BE SURE NOT TO MISTAKE THIS POISONOUS MUSHROOM FOR THE EDIBLE HONEY FUNGUS (*ARMILLARIA MELLEA*) (SEE PAGE 37) OR VELVET SHANK (*FLAMMULINA VELUTIPES*) (SEE PAGE 60).

Inocybe erubescens (Inocybe patouillardi)
Red Staining Inocybe

DEADLY POISONOUS

SEASON CHECK
July to November
(Summer to late autumn)

HABITAT
Along paths in mixed woodland, especially with beech and Sweet Chestnut, on chalk soil, and on heathland.

SIZE
Cap 2.5–8cm across. Stem 3–10cm high, 1–2cm thick.

MAIN IDENTIFICATION FEATURES
Cream-coloured cap with clear red stains on both cap and stem. Cap edges often uneven and cracked.

It is best to avoid all members of the Inocybe family. This one is the most deadly – and has on occasions been mistaken for a Chanterelle (*Cantharellus cibarius*) (see page 50). Just one (accidentally included in a dish) killed five soldiers on a survival exercise in a Scandinavian forest.

Some Inocybes are small and dull in colour, and have conical caps with rims that often become uneven, racked, lobed or split. The Red Staining Inocybe's cap is slightly conical, and ivory-cream in colour when young, staining red or brown. The gills are pink at first, darkening to light brown and bruising red. The thickish white stem is slightly bulbous at the base and marked with red stains. The flesh is white.

POISONING EFFECT
Attacks the liver and kidneys with fatal results.

WARNING:

THIS MUSHROOM GROWS AT THE SAME TIME AS THE CHANTERELLE (*CANTHARELLUS CIBARIUS*) (SEE PAGE 50). IT CAN BE MISTAKEN FOR THAT AND THE WINTER CHANTERELLE (*C. INFUNDIBULIFORMIS*) (SEE PAGE 52).

Lactarius torminosus
Woolly Milk Cap

Lactarius pubescens
Var. Woolly Milk Cap

POISONOUS

SEASON CHECK
July to November
(Summer to late autumn)

HABITAT
Widespread, often near Silver Birch on sandy soil, or along the edges of coniferous wood.

SIZE
L. torminosus: Cap 4–12cm across. Stem 4–8cm high, 1–2cm thick.
L. Pubescens: Cap 4–10cm across. Stem 3–6cm high, 1.2–3cm thick.

MAIN IDENTIFICATION FEATURES
Pale orange cap (*L. pubescens* has a white cap) with rolled rim and marked woolly edge.

Var. Woolly Milk Cap

Except in their colour, these two mushrooms are identical, and both are poisonous. *L. torminosus* has a similar colour to the much-prized Saffron Milk Cap (*Lactarius deliciosus*) (see page 70), so could be mistaken for it. *L. torminosus* and the Saffron Milk Cap also grow in similar places, so be very careful.

The caps of *L. torminosus* are convex, becoming funnel-shaped with hairy, in-rolled rims. They are buff to deep pink in colour, with deeper concentric bands. The gills are narrow and pinkish, the stems pale salmony pink, slightly downy and becoming hollow. The flesh is white, with white milky juice.

Lactarius pubescens is smaller and paler than *L. torminosus* (and more rare). Its cap is cream to rosy buff, often growing even paler in the sun. It is initially convex, then centrally depressed. The rim is in-rolled and woolly, the crowded whitish pink gills darken with age. The stem is the same colour as the cap, with a rosy buff band at the top. The flesh and the milky juice (the mushroom's spores) are white.

POISONING EFFECT
Severe stomach cramps and sickness, but, thankfully, the effects are not fatal.

Woolly Milk Cap

WARNING:

DO NOT CONFUSE WITH THE SAFFRON MILK CAP (*LACTARIUS DELICIOSUS*) (SEE PAGE 70) AND VAR. SAFFRON MILK CAP (*LACTARIUS DETERRIMUS*) SEE PAGE 69).

Lepiota cristata

POISONOUS

SEASON CHECK
July to October
(Summer to autumn)

HABITAT
Mixed woodland along woodland paths. Often appears where there is garden refuse and in leaf litter generally.

SIZE
Cap 2–5cm across. Stem 2–3cm high, 0.3–0.4cm thick.

MAIN IDENTIFICATION FEATURES
Bell-shaped cap with reddish-brown markings and red-brown centre.

This is one of the Lepiotas to avoid. It is much smaller than the Parasol mushroom (*Macrolepiota procera*) (see page 82), which is highly prized, but it comes from the same family and has many similar features. There are some 59 different species of Lepiota to be found in Europe, but it is best to stick to the Parasol and the Shaggy Parasol (*Macrolepiota rhacodes*) (see page 84) – they are the most edible.

L. Cristata has an irregular bell-shaped cap with reddish brown scales and a marked central reddish brown centre. The stem is flesh-coloured and has a ring. The gills are white, browning with age. The flesh is thin and white. The aroma is slight but unpleasant.

POISONING EFFECT
Non-fatal gastric upsets.

WARNING:

DO NOT CONFUSE WITH THE PARASOL MUSHROOM (*MACROLEPIOTA PROCERA*) (SEE PAGE 82) OR THE SHAGGY PARASOL (*M. RHACODES*) (SEE PAGE 84).

Mycena pura

POISONOUS

SEASON CHECK
July to November
(Summer to late autumn)

HABITAT
Usually in leaf litter under beech or birch.

SIZE
Cap 2–5cm across. Stem 5–10cm high, 0.4–1cm thick.

MAIN IDENTIFICATION FEATURES
Bell-shaped pink to lilac cap. Tends to crowd together in troops.

Some books refer to this mushroom as edible, but in fact it has hallucinatory effects and should be avoided. There is sometimes confusion between this mushroom and the Amethyst Deceiver (*Laccaria amethystina*) (see page 67), which grows at the same time and in similar locations.

The cap is convex in shape and variable in colour from lilac to pink, lined at the margin when wet, but paler when dry. The gills are pink. The stem flushed pink, with fine white fibres at its base. The flesh is white. The aroma is slightly of radishes.

WARNING:

BE CAREFUL NOT TO CONFUSE THE AMETHYST DECEIVER (*LACCARIA AMETHYSTINA*) (SEE PAGE 67) WITH THIS MUSHROOM.

POISONING EFFECT
Can cause hallucinations, which may be disturbing.

Omphalotus olearius

POISONOUS

SEASON CHECK
August to November
(Late summer to late autumn)

HABITAT
The bases or stumps of trees – usually oaks, chestnuts and olive trees (which are also the trees favoured by Chanterelles).

SIZE
Cap 5–10cm across. Stem 4–14cm high, 0.7–3cm thick.

MAIN IDENTIFICATION FEATURES
Bright orange caps with tapered stems. Grows in clusters on tree bases or roots.

This mushroom is included as it can be confused with the Chanterelle (*Cantharellus cibarius*) (see page 50), which grows in similar places. Fortunately *Omphalotus olearius* is quite rare, but it is worth noting.

The cap is bright orange and almost funnel-shaped with irregular edges. The stem is a paler orange, wavy and markedly tapering towards the base. The flesh is yellowish, darkening towards the stem base. These mushrooms grow in clusters at the base of trees, on the roots or on decaying stumps. They can sometimes be seen glowing in the dark, due to phosphorescence coming from the gills. The aroma is unpleasant: a clear indication that it is not a Chanterelle.

WARNING:

DO NOT CONFUSE WITH A CHANTERELLE (*CANTHARELLUS CIBARIUS*) (SEE PAGE 50).

POISONING EFFECT
Unpleasant symptoms including stomach cramps and sickness.

Paxillus involutus
Brown Roll Rim

POISONOUS, POTENTIALLY DEADLY POISONOUS

SEASON CHECK
Late July to late November (Summer to late autumn)

HABITAT
Usually associated with broad-leaved trees on acidic soil, often along path edges under birch.

SIZE
Cap 5–12cm across. Stem 7.5–8cm high, 1.2cm thick.

MAIN IDENTIFICATION FEATURES
Marked rusty brown in-rolled cap with a viscid appearance.

This mushroom is very common and must be carefully identified, especially when young, as it has been mistaken for the Chanterelle (*Cantharellus cibarius*) (see page 50). It can resemble *Boletus badius* from the top, although on closer inspection it has gills not pores. The cap is rust brown with a shiny surface when wet, with a slightly woolly appearance to the rim. When the species is young the rim is in-rolled – hence the name. As it grows, however, the in-rolled rim becomes less evident. This mushroom was, until recently, often shown in some French books as edible after treatment. This is now known not to be the case as the toxin builds up in the body and you can eventually die after some years of ingestion.

POISONING EFFECT
Deadly if eaten raw. Even when cooked, the toxins remain, causing damage to the liver and kidneys, which can be fatal.

WARNING:

DO NOT CONFUSE WITH A CHANTERELLE (CANTHARELLUS CIBARIUS) (SEE PAGE 50).

Pholiota squarrosa
Shaggy Pholiota

POISONOUS

SEASON CHECK
September to November (Autumn)

HABITAT
The base of deciduous trees, but sometimes also on conifers.

SIZE
Cap 3–10cm across. Stem 5–12 cm high, 1–1.5cm thick.

MAIN IDENTIFICATION FEATURES
Shaggy yellow cap and stem. Grows in clusters.

This toxic mushroom has been mistaken for some edible species, and you should know it in order to avoid such confusion. There are several others in this family: it is best to leave them all alone.

The cap is straw-coloured and covered with up-turned reddish scales. It is convex to begin with, then flattens out. The margin remains in-rolled. The pale yellow stem is also scaly and it has a clear ring below the cap. The flesh is tough and yellow, becoming reddish brown at the base. These mushrooms grow in clusters with several joined together and are quite common in the autumn. The smell is of radishes.

WARNING:

AVOID THIS: IT IS INEDIBLE AND UNPLEASANT – DO NOT CONFUSE WITH ANY EDIBLE SPECIES.

POISONING EFFECT
Unpleasant symptoms following ingestion of the toxins, including sickness and diarrhoea.

Russula emetica
The Sickener

Russula nobilis (R. mairei)
The Beechwood Sickener

POISONOUS

SEASON CHECK
July to November
(Summer to late autumn)

HABITAT
Almost exclusively under pine (Beechwood Sickener almost exclusively under beech).

SIZE
Cap 3–10cm across. Stem 4–9cm high, 0.7–2cm thick.

MAIN IDENTIFICATION FEATURES
Vivid red cap, sometimes marked with whitish blotches. White stem.

The Sickener

There are 112 mushrooms in the Russula group. Some are very edible, but you should avoid any with red caps. These include the Sickener and the Beechwood Sickener, which are very similar in appearance and equally poisonous but found in different habitats.

Russula Emetica's cap is cup-shaped at first, flattening later with a shallow central depression. The striking and distinctive colour – scarlet, cherry or blood red – may be marked by faded whitish patches. This mushroom peels easily to reveal reddish, crumbly flesh beneath, which can also be rather sticky. The widely-spaced white gills darken with age, the stem is white, cylindrical and sometimes slightly swollen at the base. There is a sweet fruity aroma.

POISONING EFFECTS
As its name suggests, ingestion will cause sickness. Fortunately, however, the effects are not fatal.

The Beechwood Sickener

WARNING:

AVOID ANY RUSSULA WITH A RED CAP.

Scleroderma citrinum
Common Earth-ball

POISONOUS

SEASON CHECK
July to November
(Summer to late autumn)

HABITAT
Mossy or peaty ground, often on heathland and in mixed woodland. Also likes sandy soil.

SIZE
Fruit body 2–12cm across.

MAIN IDENTIFICATION FEATURES
Appearance of a yellowish warty ball. Often grows in large numbers.

This is one of the few Puff-ball-type mushrooms that is not edible. Unfortunately it is very common.

The fruit body is attached to the soil by small threads. The ball has a warty appearance and is dirty yellow to light brown in colour. The size varies widely. Cut through, it usually reveals the soil-like spore. They can be lemon yellow when young, so be careful with identification. Several stages of growth are photographed here.

POISONING EFFECT
Unpleasant, but non-fatal symptoms, including sickness and stomach cramps.

WARNING:

THIS HAS A STRONG, REVOLTING SMELL – LIKE THE RUBBER OF OLD CAR TYRES – AND IS NOT EDIBLE: AVOID IT.

Stropharia aeruginosa
Verdigris Agaric

POISONOUS

SEASON CHECK
June to November
(Early summer to late autumn)

HABITAT
Widespread on heathland, along paths and open mixed woodland.

SIZE
Cap 2–8cm across. Stem 4–10cm high, 0.4–1.2cm thick.

MAIN IDENTIFICATION FEATURES
Verdigris-coloured cap, flecked with scales. Blue-white stem with scaly or hairy base.

This poisonous mushroom may be confused with the Aniseed Toadstool (*Clitocybe odora*) (see page 55): they often grow in similar locations and at the same time.

The glutinous mould green cap is convex to bell-shaped, then flattens out. It is flecked with white scales, which turn yellow as the gluten disappears. The gills begin white, then turn brown. The stem is dirty bluish white with whitish fibrous hairs or scales above a bulbous base. The flesh is bluish white, darker at the base.

WARNING:

LEAVE THIS TOXIC MUSHROOM ALONE – IT WILL MAKE YOU ILL IF YOU EAT IT.

POISONING EFFECT
Unpleasant symptoms following ingestion of the toxins, including sickness and diarrhoea.

Tylopilus felleus
Bitter Boletus

DANGEROUS

SEASON CHECK
August to November
(Late summer to late autumn)

HABITAT
Deciduous and coniferous woodland.
Often found in small groups.

SIZE
Cap 6–15cm across. Stem 7–10cm high, 2–3cm thick (but often up to 6cm at the base).

MAIN IDENTIFICATION FEATURES
Light brown cap with clearly marked network on the stem and pinkish pores.

You will often see this mushroom's cap and think you have found a Cep (*Boletus edulis*) (see page 44). Alas, closer inspection will reveal that you have found the Bitter Boletus – well named for its horrible bitter taste. Include even one and your mushroom dish will be ruined.

The cap is snuff brown and a little downy at first, then becoming smooth. The pores are pinkish in colour, discolouring with age. The large bulbous stem, yellowish in colour, has a marked brownish network towards the base. The bitter aroma is a tell-tale sign of this mushroom, which often grows to a large size.

WARNING:

BEWARE OF THIS MUSHROOM, AS ALTHOUGH IT IS EDIBLE, IT WILL SPOIL ANY DISH IT IS INCLUDED IN BECAUSE OF ITS REPELLENTLY BITTER TASTE.

STORAGE METHODS

The art of preserving food is as old as the hills. Since mushrooms are very seasonal, a number of methods have been developed to enable us to eat them all the year round.

On the mushroom identification pages, I have mentioned the best storage methods recommended for each type of mushroom. However, there are other specific methods worth considering as well. Here's an introduction to several you can try.

Drying

The most important way of preserving mushrooms is to dry them over a gentle heat or in the open air. This method has been used for many years, and dried wild mushrooms are now a very big business worldwide. There can be a substantial difference in the quality sometimes, so be careful what you buy. When air-drying it is not possible to control the humidity: sometimes the mushrooms are not dry enough, which can lead to a spoilt batch (if you are not careful).

I have found the best method is to use a dehydrator – a machine especially designed for drying food. Dehydrators are now readily available, and I have included a reference to some contacts in the *Useful Addresses* section (see page 153). Only use perfect mushroom specimens for drying in this way. Clean and slice them, and place the slices on trays. Set the thermostat in accordance with the manufacturer's instructions and leave them overnight.

Over the years, I have also tried drying mushrooms in airing-cupboards and over the top of an Aga, with quite good results – but dehydrators give perfect results every time. I do sometimes dry Morels on strings in the kitchen: they are the only mushrooms I dry in this way, and it is a very effective method.

Remember to store your dried mushrooms in an airtight container, or, if you prefer, freeze them (see page 146). I often use airtight glass jars: they look rather attractive in the kitchen. Indeed, when we had our pub-restaurant we would put sealed glass jars of mushrooms on shelves around the lounge bar, which led to it being dubbed 'the mushroom pub'.

When you dry mushrooms, you will usually end up with a few broken-off fragments as well. Don't throw these bits away – they can be made into powder, using either a pestle and mortar or a liquidiser, and stored in airtight containers. This mushroom powder can be added directly to soups and sauces to give extra flavour – it's particularly good in pasta sauces.

When you are using dried mushrooms, always remember to soak them in plenty of luke-warm water for at least 25 minutes before straining them. Keep the water, too: it can be used for making a sauce as a thin stock to cook the rice in if you are making a risotto.

You can use dried mushrooms directly in stews and other slow cooking dishes, as they will reconstitute during the cooking, but do remember they have an intense flavour, so use them sparingly.

Salting

Salting mushrooms to preserve them has been customary in Eastern Europe for many years, employing the same simple method that is often used for fish and meat. Choose only perfect mushrooms, and beware of any with maggot holes. Clean the mushrooms well by brushing them (rather than washing in water) and slice the larger ones. Now arrange them in alternate layers of salt and mushrooms, allowing approximately 60g of sea or rock salt per 1kg of fresh mushrooms. The salt will dissolve into a preserving brine, and the mushrooms should then be stored in sealed jars or containers. When you come to use them, always remember to rinse them well with fresh water first and then dry them. I have used this storage method very successfully with Saffron Milk Caps (*Lactarius deliciosus*) (see page 70).

Freezing

Freezing is another method for storing, but I have found that this is generally unsuccessful – except with mushrooms from the Blewit family, which do freeze well. Restaurants often store sliced wild mushrooms in specially constructed freezers, in trays that avoid damage to the mushrooms, but this equipment is expensive and not sensible for the home.

I have often saved mushrooms in butter to use for sauces. This preserves their texture and taste and makes sauce-making easy. Quickly fry the mushrooms in butter and then freeze them in blocks, which can be stored in containers. You must use plenty of butter and remember to label and date the batches. When you want to use them, thaw them at room temperature first. This method is very successful with Chanterelles (*Cantharellus cibarius*) (see page 50) and Ceps (*Boletus edulis*) (see page 44). I also often make duxelles of mushroom sauces, freezing them in ice trays for later use in soups and stews.

Pickling

Pickling is another method of storage, which I have used to good effect. Again, only use perfect specimens. Clean them well and blanch them in lightly salted water, then drain and dry them. Heat up the olive oil or vinegar (whichever you prefer). Put the mushrooms in a sterilised jar and top up with the hot liquor, then seal the jar and store it. Although oil is more expensive, it has an added advantage: when you open up the jar, you have not only the mushrooms but mushroom oil too – excellent for salad dressings!

Sometimes a season may bring a glut of Horse or Field Mushrooms (*Agaricus arensis/A.campestris*) (see page 28 and 30). When this happens, I often make a batch of mushroom ketchup. Clean the mushrooms and put them in a pan with plenty of water and a little salt. Bring to a good rolling boil for 20 minutes, then sieve the liquor, press out all the juice, and discard the spent mushrooms. Return the juice to a pan and reduce by half, then bottle the ketchup for use later, storing it in the refrigerator to avoid fermentation. And remember, it's strong, so use it sparingly!

I have successfully stored some of the sweeter and more colourful mushrooms in spiced alcohol (I would especially recommend vermouth and white rum). You need to blanch the mushrooms first, before drying them and putting them into a jar. Have the spiced alcohol ready-heated (but not boiled) and pour it over the mushrooms until they are completely covered.

Then seal the jar and store for at least a month before use. This makes a great sauce to go with ice cream. Delicious!

COOKING WITH MUSHROOMS

I have been cooking with wild mushrooms for more than four decades, and their diverse flavours and tastes never cease to amaze me. In recent years they have become a significant part of a good chef's repertoire, and we have all enjoyed the results. There is no doubt that the French and Italians have done more than anyone to make us more aware of the great culinary usefulness of mushrooms. There are now very few good restaurants that do not have wild mushrooms on the menu.

On the mushroom identification pages I have given cooking suggestions for each individual species of mushrooms. Their tastes and textures vary widely. It's important to have a good idea what dishes each species works best with.

Wild mushrooms are rich, so it is best not to overdo the quantities you eat at any one meal. It's also important to remember that some people are allergic to some of the edible mushrooms: so serve with care!

When teaching at cookery schools I have found it effective to let the students try out each mushroom. In this way they learn the individual flavours for themselves, and can decide what they prefer. Surprisingly, at one school in Scotland, the Cauliflower Fungus (*Sparassis crispa*) (see page 104), cooked in a light tempura batter, scored the best results.

The simplest way of cooking mushrooms is, in my opinion, the best. Fry them fiercely in a little butter and oil for a few minutes. Then take them out of the pan and put the remaining juice aside for later use in a sauce. Return the mushrooms to the pan with a little butter, and, once the butter has melted, serve them immediately. Perfect!

You can sauté mushrooms using a lower heat, this method is best if you are using both fresh and dried in your dish. Grilling can be used for the bigger mushrooms but always use a good drizzle of oil to stop them drying out.

Some species must be blanched before cooking (this is also a good way to preserve the more delicate, smaller mushroom). Do make sure that you always discard the blanching water.

You can deep-fry mushrooms and this method is very good for a tempura batter. You can also dip them in seasoned flour then beaten egg and breadcrumbs before deep-frying. This is an excellent way of serving a mixture of the more solid mushrooms, and they tend to go well with a sweet chilli dipping sauce.

I only use the microwave for reheating made-up dishes.

Wild mushrooms make an ideal starter to a meal. Whip up a soup, or put them in a small omelette. They also go very well with scrambled eggs: be decadent and serve scrambled eggs with shaved truffles at your next dinner party. You can also make a fine mushroom pâté, thickened with hard-boiled egg yolks.

The intense flavour of some mushrooms goes best with the stronger meats and game: members of the Morel family (see pages 87–90) and Horse mushrooms (*Agaricus arvensis*) (see page 28) are ideal for this purpose.

The Chicken of the Woods (*Laetiporus sulphureus*) (see page 72) has a very different flavour and texture. It is a fine substitute for chicken in a curry or chicken *à la crème* recipe. Ideal for vegetarians – but you will have to work to convince them that it really isn't chicken!

The more delicate meats like lamb and pork go very well with Chanterelles (*Cantharellus cibarius*) (see page 50), Saffron Milk Caps (*Lactarius deliciosus*) (see page 70) and Hedgehog mushrooms (*Hydnum repandum*) (see page 62).

Chanterelles and Saffron Milk Caps also go very well with fish. So does the Horn of Plenty (*Craterellus cornucopioides*) (see page 58). These give a dish colour as well as flavour.

Wild mushrooms are, of course, a source of delight to vegetarians, as they provide a great deal of variety as well as being a source of the essential minerals potassium, magnesium and iron. They are also rich in

Store cupboard essentials:

There are a number of items it is good to have in the store cupboard when cooking mushrooms. Over the years I have found the following of great help to get the best results out of the mushrooms:

1. A good quality olive oil (but not extra virgin as the taste is too strong).
2. Both hazelnut oil and walnut oil give added flavour to the mushrooms with their wonderfully nutty taste.
3. Many of the dishes require garlic – please use it fresh.
4. Quality fresh herbs are essential to enhance all dishes.
5. Good quality truffle oil, a few drops go a long way.
6. *Marigold Swiss Vegetable Bouillon* powder.
7. Tempura batter mix.

These will help you to get the best results from your finds.

niacin, and contain other B group vitamins. Mushrooms contain up to 8 per cent protein and are only 35 calories per 100g – but if you are calorie-counting, watch out for the butter (use olive oil instead).

Over the years efforts have been made to cultivate the more exotic mushrooms. These have been successful with some species and cultivated species are always good to consider adding to your own collected mushrooms. Since the last war, both the USA and Japan have made strenuous efforts and now produce a good range: these, added to those produced by successful European growers, give us a huge range of choice. Sophisticated technology coupled with good international transport get the crop to the table of the world quickly.

Wild Ceps and Chanterelles.

Give the commercially-grown species a try to add to your own finds and take the opportunity to check the flavour against those you have collected.

I hope you have enjoyed this cookery chapter, but don't be afraid to experiment, and remember there are many good cookery books that can help you. You can also check out my website – www.tastymushroompartnership.co.uk – for seasonal recipes.

So as always: *bon appetit!*

The more notable of the exotic species are:

- The Matsutake, which is grown in Japan and highly prized, commanding a very high price. The flavour is great and, although it is not widely available, it can be obtained through specialist suppliers (please see *Useful Addresses*, page 153).
- The Wood Ear, a cousin of the Jew's Ear (*Auricularia auricularia-Judea*) (see page 38), which is mainly grown in Japan. It is not, in my view, very tasty but adds texture to some dishes.
- The Enoki, which is a cousin of the Velvet Shank (*Flammulina velutipes*) (see page 60). It is small and white and tasty, with a crunchy texture. It can often be found in supermarkets.
- The Hen of the Woods (*Grifola frondosa*) (see page 61), which is now cultivated and very good, but not as good as the wild version.
- The Wood Blewit (*Lepista nuda*) (see page 76), which is now grown commercially. Much paler than its wild cousin, in my view, it has nothing like as good a flavour.
- The Shiitake, which is now very common, being grown worldwide. It has a nice flavour, a firm texture and is good to eat.
- The Oyster mushroom, which comes in a very wide range of colours not found in the wild. These are widely available and good, the added colour often being helpful with mixed dishes.

A cultivated selection from the market.

An Example of Cooking With Mushrooms

This dish is basically a fricassee of mixed mushrooms. It consists of a combination of wild mushrooms, cultivated mushrooms and some wild, as seasonably available.

My friend and fellow mushroom hunter Ian Howell, who is Head Chef at *The Swan Hotel* in Southwold, often serves wild mushrooms in this way, either on their own or as a sauce to be added to fish or meat.

This example dish is easy to prepare and highlights the wonderful flavours of the mushrooms in a simple way.

1. The mushrooms, together with finely chopped garlic and onions, are sautéed in unsalted butter over a high heat for approximately 45 seconds.

2. Chopped parsley and tomato are then added.

3. A dessert-spoon of water is then added to aid the infusion.

4. The dish is plated up and finely grated fresh parmesan added. The result is stunning!

FURTHER READING

Encyclopedia of Fungi of Britain and Europe
Michael Jordan
(Frances Lincoln)

Field Guide to Mushrooms and Other Fungi of Britain and Europe
(New Holland)

Mushrooms and Other Fungi of Great Britain and Europe
Roger Phillips
(Pan Books)

Truffles The Black Diamond and Other Kinds
Jean-Marie Rocchia
(Editions A. Barthelemy)

A good companion book is:

Food From the Wild
Ian Burrows
(New Holland)

For the cooks among you:

Mushroom Feasts
Steven Wheeler
(Southwater)

USEFUL ADDRESSES

The Tasty Mushroom Partnership
Poppy Cottage
Skilman's Hill
Southwold
Suffolk
IP18 6EY
www.tastymushroompartnership.co.uk

The Association of British Fungus Groups
Harveys
Alston
Axminster
Devon
EX13 7LG
www.abfg.org

British Mycological Society
The Wolfson Wing
Jodrell Laboratory
Royal Botanic Gardens
Kew
Richmond
Surrey
TW9 3AB
www.britmycolsoc.org.uk

EziDri (dehydrators)
116 Preston Road
Yeovil, Somerset
BA20 2DY
Tel: 01935 471 870

The following organisations often arrange forays on a local basis so are worth contacting:

Forestry Commission
www.forestry.gov.uk

The Wildlife Trusts
www.wildlifetrusts.org

GLOSSARY

Ascomycetes	Group of mushrooms, which bear their reproductive spores in a sac.
Basidiomycetes	Group of mushrooms characterised by the presence of spore-bearing cells called Basidia.
Bracket	Mushroom attached to trees like a bracket or shelf.
Cap	Part of a mushroom bearing gills or tubes.
Convex	Curved or rounded outwards.
Decurrent	Running down the stem (usually referring to gills).
Fibrous	Composed of fine fibres or threads.
Flesh	Inner tissues of a mushroom.
Fruit body	Structure which bears spore-producing cells.
In-rolled	Curled inwards and down.
Marginate	With a distinct ridge or margin.
Milk	Sticky fluid released by some mushrooms.
Network	Mesh or pattern of criss-crossed fine ridges.
Partial veil	Fine web of tissue, which connects the cap margin to the stem.
Pores	Openings of tubes in Boletus and Polypore mushrooms.
Ring	Remains of the partial veil left on the stem.
Scales	Small-to-large raised flakes on the surface of the cap or stem.
Spore print	Thick deposit of mushroom spores allowed to drop onto paper.
Stem	Stalk that carries the cap.
Striated	Well-marked parallel grooves or lines, especially at the cap edges.
Tubes	Underside cluster of tubes on Boletus and Polypore mushrooms, which contains the spores.
Universal veil	Fine-to-thick covering tissue, which envelops some immature mushrooms.
Volva	Thick universal veil remnant, which forms a bag or sheath at the base of the stem.

INDEX

Page numbers in *italic* refer to illustrations; those followed by **g** denote an entry in the glossary.

ACKNOWLEDGEMENTS

Thanks go to the following people and organisations for their help with the book:

L'Aquila for the fresh truffles, which we needed for the photographs; Frank Barratt for his wonderful pictures of Borough Market; Lennie Doy and his dog Sam for their hunting skills; Peter Henley for all his support and company on our many mushroom hunts; Michael Jordan for all his help with technical detail and support; Margaret Melicharova for her editing of my original manuscripts; Nutters of Southwold and Louise for her patience when shooting mushroom products in her shop; *The Swan Hotel*, Southwold and Head Chef, Ian Howell, for the cookery shots in his busy kitchen; all of the people on our forays this year, who have helped to find some of the more unusual mushroom that we have included in the book.

A special thanks to my wife Valerie, who has lived this book from start to finish, for her patience and support in the hunt.

Photography/Artwork Acknowledgements

All photographs taken by Peter Henley and Peter Jordan, apart from:

Yves Deneyer: 10 (*left*); 43 (*top*); 40; 46; 78; 89; 92; 98; 100; 101; 103; 114; 125; 130; 142; 143
Frank Barratt: 108, 150
Michael Jordan: 124; 137

All line artworks by Peter Henley
Cartoon artwork by Tony Hall

Opener Images

Page 5 (*clockwise from top left*):
Field Mushroom (*Agaricus campestris*); Var. Morel (*Morchella esculenta*);
St George's Mushroom (*Calocybe gambosa*); Chanterelle (*Cantharellus cibarius*);
Stinkhorn (*Phallus impudicus*); Amethyst Deceiver (*Laccaria amethystina*).

Page 27 (*clockwise from top left*):
Chanterelle (*Cantharellus cibarius*); Brown Birch Boletus (*Leccinum scabrum*);
Var. Morel (*Morchella esculenta*); Parasol (*Macrolepiota procera*); Var. Orange Birch Boletus (*Leccinum quercinum*); Var. Puff-ball (*Lycoperdon perlatum*).

Page 113 (*clockwise from top left*):
Common Ink Cap (*Coprinus atramentarius*); Woolly Milk Cap (*Lactarius torminosus*); Death Cap (*Amanita phalloides*); Brown Roll Rim (*Paxillus involutus*); The Sickener (*Russula emetica*); Destroying Angel (*Amanita virosa*).